Automated Process Control Strategies

A Special Publication

Prepared by **Task Force on Automated Process Control Strategies**
Robert Hill, *Chair*

Orris E. Albertson
Herb Campbell
James L. Daugherty
Kevin J. Deeny
Richard E. Finger
M. Truett Garrett, Jr.
Joseph J. Gemin
Robert A. Gillette
Harold D. Gilman
Theodore S. Glanton
John D. Haase
Richard S. Haugh
James A. Heidman
Agamemnon D. Koutsospyros
David O. Loehden
Miryoussef Norouzian

David Olson
Tony M. Palmer
Steven R. Reusser
Glenn A. Richter
George R. Schillinger
Walter W.M. Schuk
Sean Scuras
Brian Smythe
Michael K. Stenstrom
Joe P. Stephenson
Michael W. Sweeney
Doug Thompson
John S. Trofatter
Paul Trout
David A. Vaccari
Cello Vitasovic

Under the Direction of the
**Instrumentation, Control, and
Automation Committee**

1997

**Water Environment Federation
601 Wythe Street
Alexandria, VA 22314-1994**

IMPORTANT NOTICE

The contents of this publication are for general information only and are not intended to be a standard of the Water Environment Federation (WEF).

No reference made in this publication to any specific method, product, process, or service constitutes or implies an endorsement, recommendation, or warranty thereof by the Federation.

The Federation makes no representation or warranty of any kind, whether expressed or implied, concerning the accuracy, product, or process discussed in this publication and assumes no liability.

Anyone using this information assumes all liability arising from such use, including but not limited to infringement of any patent or patents.

Library of Congress Cataloging-in-Publication Data

Automated process control strategies / prepared by Task Force on
  Automated Process Control Strategies, under the direction of the
  Instrumentation, Control, and Automation Committee.
        p.   cm. — (Special publication)
    Includes bibliographical references (p.   ) and index.
    ISBN 1-57278-063-0 (pbk)
      1. Sewage disposal plants—Automation.   2. Process control.
  I. Water Environment Federation. Task Force on Automated Process
  Control Strategies.   II. Water Environment Federation. Instrumentation,
  Control, and Automation Committee.   III. Series: Special publication
  (Water Environment Federation)
  TD746.A98   1997
  628.3—dc21   97-838
                                                                 CIP

ISBN 1-57278-063-0
Printed in the U.S.A. **1997**

Water Environment Federation

Founded in 1928, the Water Environment Federation is a not-for-profit technical and educational organization. Its goal is to preserve and enhance the global water environment. Federation members number more than 42,000 water quality professionals and specialists from around the world, including engineers, scientists, government officials, utility and industrial managers and operators, academics, educators and students, equipment manufacturers and distributors, and other environmental specialists.

For information on membership, publications, and conferences, contact

Water Environment Federation
601 Wythe Street
Alexandria, VA 22314-1994   USA
(703) 684-2400

Special Publications of the Water Environment Federation

The WEF Technical Practice Committee (formerly the Committee on Sewage and Industrial Wastes Practice of the Federation of Sewage and Industrial Wastes Associations) was created by the Federation Board of Control (now Board of Directors) on October 11, 1941. The primary function of the Committee is to originate and produce, through appropriate subcommittees, special publications dealing with technical aspects of the broad interests of the Federation. These manuals are intended to provide background information through a review of technical practices and detailed procedures that research and experience have shown to be functional and practical.

Contents

List of Tables

List of Figures

Acknowledgments

Authors of the manual are

Orris E. Albertson
Herb Campbell
Kevin J. Deeny
Richard E. Finger
M. Truett Garrett, Jr.
Harold D. Gilman
Richard S. Haugh
James A. Heidman
Robert D. Hill
David O. Loehden
David Olson

David M. Osborn
Tony M. Palmer
Steven R. Reusser
Walter W.M. Schuk
Michael K. Stenstrom
Joe P. Stephenson
Michael Sweeney
Doug Thompson
David A. Vaccari
Cello Vitasovic

In addition to the Authors, Reviewers include

James L. Daugherty
Joseph J. Gemin
Robert A. Gillette
Theodore S. Glanton
John D. Haase
Agamemnon D. Koutsospyros
Miryoussef Norouzian

Glenn A. Richter
George R. Schillinger
Sean Scuras
Brian Smythe
John S. Trofatter
Paul Trout

Authors' and Reviewers' efforts were supported by the following organizations:

American Bottoms Regional Wastewater Treatment Facility, Sauget, Illinois
Cedar Rapids Water Pollution Control Facilities, Iowa
CH2M Hill, Englewood, Colorado
City of Indianapolis, Indianapolis
City of Houston Public Utilities, Texas
Controlotron, Hauppauge, New York
Enviro Enterprises, Inc., Salt Lake City, Utah
Enviromega, Inc., Hamilton, Ontario, Canada
Espey, Huston & Associates, Inc., Houston, Texas
Greeley and Hansen, Philadelphia, Pennsylvania
HDR Engineering, Inc., Omaha, Nebraska
Hydromantis, Inc., Hamilton, Ontario, Canada
John Carollo Engineers, Sacramento, California

Junkins Engineering, Morgantown, Pennsylvania
KC Environmental, Inc., Levittown, Pennsylvania
Madison Metropolitan Sewerage District, Wisconsin
Municipality of Metropolitan Seattle, Renton Treatment Plant, Washington
Proctor and Redfern Ltd., Kitchener, Ontario, Canada
Professional Services Group, Marshfield, Massachusetts
Reid Crowther Consulting, Inc., Seattle, Washington
Scuras Engineering Services, Boulder, Colorado
Stevens Institute of Technology, Hoboken, New Jersey
Thorn Creek Basin Sanitary District, Chicago Heights, Illinois
University of California, Los Angeles
University of New Haven, Department of Civil and Environmental
    Engineering, Connecticut
U.S. Environmental Protection Agency, Cincinnati, Ohio
Walter W. Shuk, Inc., Batavia, Ohio
Water Technology International Corp., Burlington, Ontario, Canada
Westin Engineering, Sacramento, California

Chapter 1
Introduction

DOING MORE WITH LESS

A common theme among wastewater managers and operators is "doing more with less." Doing more with less means operating and maintaining the wastewater infrastructure at the same or a higher level of performance with a smaller budget. Doing more with less means working with fewer people on staff, reducing power consumption, and decreasing chemical consumption. Many professionals consider doing more with less to be a change of paradigms within the industry.

What social, economic, and technological forces have moved the wastewater treatment industry in this direction? Competition has had a significant effect. Competition takes the form of allocating limited budget funds among the water and wastewater utilities, police and fire protection, road and bridge maintenance, health and social services, and all of the other functions of local government. With few exceptions, governments do not have enough budget money to meet all the perceived needs. In many cases, the total budget has decreased leading to even more competition among the individual departments. The wastewater industry may even be a victim of its own success if the public no longer feels the necessity to allocate additional resources to this function.

Competition also takes the form of private industry. Several national and international companies now manage, operate, and maintain public wastewater treatment plants (WWTPs). In one recent Ohio case, a private company even purchased a WWTP from the local government. Clearly, government no longer has a monopoly on providing wastewater treatment services. While it is not within the scope of this manual to debate the benefits and dangers of privatization, the resulting competition certainly is a driving force in the wastewater treatment industry.

Leading utilities, both public and private, are now using the strategic application of technology to gain a competitive advantage. They are reengineering and applying technology to almost every business process of the utility including operations, maintenance, engineering, customer service, laboratory practices, and administration. One of the most cost effective of these major technologies is process control. Correct application of process control has proven in numerous cases to improve efficiency and productivity of a utility while reducing costs. Probably the most important aspect of process control is designing what process control strategies will most advantageously affect the organization.

CONTROL STRATEGIES

This manual presents several different control strategies, algorithms, and objectives for many of the common unit operations in WWTPs. While these ideas should prove useful to design and operations engineers, this manual cannot be considered a ''cookbook'' design manual. These strategies have been proposed by a large number of individuals with widely varying operational philosophies. Some strategies may be partially or even completely incompatible with others. For instance, one strategy may require a parameter to be increased under certain conditions, while another strategy will decrease that same parameter. Certainly, there is as much diversity in control strategies as there is in design philosophies.

Some strategies have been used for many years and are well documented with one or more full-scale references. Other strategies have not been widely applied at full scale. These strategies often have only been documented at a theoretical level or at pilot scale. While the authors have excluded several control strategies on which the committee could not reach consensus, many other underdocumented strategies have been retained because of their potential benefits. The reader is cautioned always to use good engineering judgment in applying any of the control strategies.

STARTING DOCUMENT—THE PROCESS AND INSTRUMENT DIAGRAM

A well-operating WWTP almost always starts with an excellent design. Likewise, an excellent design is almost always the result of consistency across the entire system, reflecting a consensus and cooperation among the process, the structural, the mechanical, the civil, the electrical, and the instrumentation and control designers. None of these groups should work in isolation from the others.

An excellent tool and starting point for any design is the process and instrumentation diagram (see Chapter 4). A process and instrumentation diagram is a schematic diagram showing the major components of a process or processes, interconnections among them, instruments, measuring points, and control elements. If the designers can reach a consensus on these major design components, much of the rest of the design is straightforward and simplified. Because of the importance of the process and instrument diagram to the design phase of each project, this manual includes at least a simplified process and instrument diagram for each unit operation to illustrate its use.

WHAT IS INCLUDED AND NOT INCLUDED IN THIS MANUAL

This manual is intended as a design guide to be used by environmental engineers performing WWTP design and instrumentation and control engineers designing and implementing control systems at WWTPs. It will also be useful to WWTP managers, operating engineers, and any other person operating or maintaining wastewater processes that have control systems.

The content of this manual was specifically restricted to control strategies and algorithms for wastewater treatment processes. The Water Environment Federation manual *Instrumentation in Wastewater Treatment Facilities* (WEF, 1993) discusses the selection of instruments, the installation details, the sizing of final control elements, and the selection of controllers and computers. Additionally, several technical organizations exist to help designers. The Instrument Society of America (ISA) provides guidelines for virtually all aspects of instrumentation. The ISA also provides excellent training and educational materials. The Instrumentation Testing Association provides detailed results of full-scale testing of instruments for water and wastewater conditions.

This manual is organized by unit operations, roughly moving from the beginning of a typical WWTP through the final discharge. Both liquid treatment and solids treatment processes are included, although the level of detail for solids treatment processes is less than that for liquid treatment processes. This discrepancy reflects the different levels of sophistication in the present state of the art. However, research is presently underway to improve instrumentation and control in the area of solids treatment processes.

Each section of this manual presents a simple description of one unit operation with an emphasis on elements that can be controlled. One or more control strategies are presented that are useful for improving process performance, reducing costs, and/or reducing maintenance. A process and instrument diagram of each process is presented. A list of minimum instrumentation required for these strategies is presented.

SOURCES OF ADDITIONAL REFERENCE MATERIAL

Instrument Society of America
67 Alexander Drive
P.O. Box 12277
Research Triangle Park, N.C. 27709

Water and Wastewater Instrumentation Testing
Association of North America
1401 I Street, NW, Suite 900
Washington, D.C. 20005

REFERENCE

Water Environment Federation (1993) *Instrumentation in Wastewater Treatment Facilities*. Manual of Practice No. 21, Alexandria, Va.

Chapter 2
Why Is Automatic Control Needed?

*I*NTRODUCTION

The owners and operators of wastewater treatment plants (WWTPs) are committed to meeting state and federal permit requirements with the use of reasonable and acceptable amounts of energy, chemicals, equipment, and operator hours. An acceptable level of resource use is established through the preparation and approval of an operating budget. Additional goals may be set to minimize waste, but the processes must be properly controlled to maintain permit compliance. The risks of operating WWTPs now include fines and incarceration for permit violations as well as public outrage at plant odors and unsightly or unhealthy conditions in the receiving stream. Automated process controls along with modern instruments provide a means to reduce the risk of permit violations and to control the economics of the operation. Awareness of the risk must be maintained, and supplies and procedures must be kept in place to allow the prompt repair of break-downs.

In past years, the cost of providing automatic control was often offset by the cost of 24-hour staffing or of extra energy or chemical use. Activated-sludge WWTPs are operated unattended with sufficient aeration and auto-mated chemical feed to meet the demands of the peak loads. This strategy of operation has been satisfactory for compliance with permit limits for

---

biochemical oxygen demand and suspended solids because excess dissolved oxygen is not objectionable. However, this approach no longer applies to effluent chlorination. The new requirements for dechlorination and for performance of biomonitoring tests without final laboratory dechlorination have made it essential to provide control through 24-hour operator staffing or through automatic control of chlorination and dechlorination where this method of disinfection is used.

For medium-sized WWTPs—those large enough to require dechlorination and biomonitoring but small enough to have been previously staffed for one shift per day—the cost of control for dechlorination is the cost of an additional 14 staffed shifts per week—possibly $80 000 per year. There is an opportunity to reduce the cost of control through the use of on-line chemical analysis and automatic control of both chlorine and dechlorination feed rates. Automation also provides the potential to reduce chemical costs through the more efficient control that results.

The cost of energy for aeration accounts for most of the electrical cost in activated-sludge WWTPs. Even though the savings in electrical energy are in the lower priced, high-consumption bracket, careful control may also provide a reduction in the demand charge or shift consumption during the off-peak hours of the day. Automated blower control at a 1.58-m^3/s (36-mgd) activated-sludge WWTP in Houston resulted in a 20% reduction in plant electricity cost (Garrett, 1990).

Today, automatic control involves the use of loop and logic control of motors and motor-operated gates and valves in response to sensors, plus data logging, alarms, and report generation. Sometimes, operations are initiated by a remote manual signal to test a strategy before making the commitment to fully automatic control. There is a wide selection of equipment available, as discussed in the chapters that follow. To achieve maximum savings of operator and technical staff time, as well as savings of energy and chemicals, computer control system designs for municipal WWTPs should be able to accomplish at least two basic functions:

- Perform those operations that an operator would typically do. Examples are the following: log all data required for a daily operation report and store the data in a database for later evaluation and use by file transfer; provide graphical presentation of the trends of on-line and archived data; and prepare and print the daily, monthly, and other reports as required.
- In cooperation with an experienced operator and/or process control engineer, provide improvements over manual control and/or provide automatic controls to reduce operator attendance. This may include process control loops to improve chemical (including oxygen) dosing, starting and stopping equipment based on process demands, placing process units on or off line, restarting equipment after power

interruption, and using expert systems or process optimization software to fine-tune the process. The control system variables, such as loop setpoint and tuning constants, should be logged, as well as the process variables.

DISTURBANCES

The objective of automated process control is to correct for varying conditions and for disturbances that have the potential to degrade the process performance. In wastewater treatment, conditions are always varying because the rate of flow and composition of wastewater are always changing, so "steady state" only applies to average conditions over a relatively long time scale. Control engineers that come from other disciplines may not be fully aware of the variability of wastewater. Some industrial processing wastewater, and even potable water production, does not have the variability of flow and composition that exists with municipal wastewater treatment. In addition to the effect of diurnal variation in human activity on wastewater flow and composition, other uncontrollable variations include seasonal temperature changes, industrial activity that may produce intermittent slug loads or seasonal loads, fuel spills that leak into the collection system, and flow variations from rain or snow melt. There are other variations within a WWTP. Some may be correctable; others must be accommodated by the control system.

Flow variations may be created by the following:

- Intermittent bar screen operation in a gravity flow WWTP,
- On–off operation of influent pumps,
- Recycle flows,
- Waves at the level-sensing point for flow measurement, and
- Bringing process units on line or taking them off line.

Other variations result from

- Variations in chemical strength when supply tanks are changed.
- Degradation of strength of chemicals with time; for example, sodium hypochlorite in solution slowly decomposes to salt and oxygen.
- Variations in solids concentration to be chemically conditioned for dewatering by belt press or centrifuge.
- Variations in ammonia and nitrite concentrations affecting effluent chlorine demand.
- Biochemical oxygen demand and NH_3–N slug loads from anaerobic digester supernatant.

- Disturbances induced by other process control loops.
- Equipment malfunctions.
- Operator error.

Variations may be divided into two groups: (1) variations at a frequency too high for a response, called *noise*; and (2) variations that represent a load change and require a response to prevent a process upset, called *disturbances*. The reasons to remove noise are to respond to the real load changes and to prevent damage to equipment because process variable signal variations pass through a controller to affect the output signal and the pump or valve it controls. Noise is removed by taking an average of the signal values over the cycle time of the noise. This is called *filtering*. In this case, it is a *low-pass filter* because low-frequency variations will pass through. It may be performed electronically on the analog signal (with capacitance, resistance, and inductance), by program on the digital signal (by calculating some form of moving average), or mechanically on the original sample (with a composite sampler).

*M*EASUREMENT FREQUENCY

Because of the traditional practice of sampling processes at 2- or 4-hour intervals, there is little awareness of the higher frequency variations in the concentration of the constituents, although flow variations are seen on a chart recording. Because of lack of experience with this information, there is uncertainty regarding whether to filter it or to respond to the variation as a load change. Frequencies greater than 1 Hz are typically filtered by a low-pass analog filter (Bibbero, 1977) in analog measurement or transmission operations (see Signal Conditioners, in the Instrument Society of America's *Directory of Instrumentation* [1996]). Lower frequencies may be addressed by digital filtering techniques such as moving averages (Olsson and Piani, 1992). The cause of disturbances should be investigated to see if they can be corrected physically—for example, through the addition of a stilling well to eliminate waves when measuring liquid level, by modifying pump sequencing, or by modifying operating levels. This is an important consideration in the design of WWTPs that will have automated control of the processes. The following are examples of physical corrections:

- A thickener scraper mechanism mounted on a pier in the center of the tank causes cyclic variations in the suspended solids concentration as the scraper discharges heavy solids to the sump each half revolution. An example of this is shown in Figure 2.1. These data were obtained from an activated-sludge clarifier with a diameter of 30 m (100 ft). These variations may be eliminated by suspending the

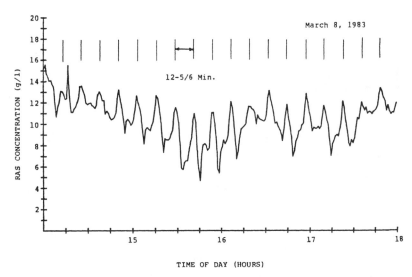

Figure 2.1   Period peaks in return activated sludge concentration (WEF, 1993)

scraper mechanism of the thickener from a bridge and putting the withdrawal point in the center of the tank.

- Constant-speed pumps cause flow pulsations as they cycle on and off to match the incoming flow, whereas Archimedes screw pumps automatically follow the incoming flow. Variable-speed pumps with automatic control of speed offer similar matching of the incoming flow except for the flow disturbance when a pump is added or taken off line.

The analog-to-digital conversion of digital controllers is typically a sample-and-hold operation. An instantaneous reading is taken and held until the next sample time. This can give values that are greatly different from those obtained with a meter having an indicating needle. Because of inertia, the needle does not respond to high-frequency noise in a direct current signal; that is, the signal reverses before the needle has traveled visibly from the average position. For process control, the noise should be removed before such a signal is used as a process variable. Otherwise, the noise will pass through the controller and create a noisy output signal. Because the frequencies and amplitude of the noise are unknown, a passive analog low-pass filter on the input to the controller is useful (Olsson and Piani, 1992). If the loop control is to be calculated at some interval longer than one second, the noise may be further attenuated by a digital filter. For this, a moving average is the simplest form of low-pass filter. While filtering is important for controllers, it should be mentioned that the unfiltered signal should be applied to some forms of stochastic models such as auto-regressive moving average (ARMA) models. Such models do their own filtering.

*A*LIASING

The sampling frequency of the digital controller should be such that the continuous signal can be reconstructed from the sampled signal. In wastewater treatment, there is seldom concern for processes having high-frequency signals that need to be followed rather than filtered. However, it is of value to be aware of the relationship between signal frequency and sample frequency.

Disturbances and process states must be sampled at a frequency sufficiently fast to detect any true variations—at least twice as fast as the maximum frequency of interest. This minimum sampling frequency is known as the *Nyquist frequency*. If a signal is sampled at a rate slower than the Nyquist frequency, the reconstructed (estimated) signal will contain frequencies and variations that do not appear in the real signal. Such effects are known as *aliasing*. *Aliasing* is a term applied to the result of periodic sampling of a varying signal and an attempt to reconstruct the signal from the sample values. Sampling theory considers that an infinite time is available to reconstruct the signal. If the sample frequency is less than twice the signal frequency, the reconstructed signal has a frequency less than the original signal. In practice, sampling at 10 times the signal frequency is commonly used to reconstruct the signal (Bibbero, 1977). A low-pass analog filter is also known as an *anti-aliasing filter*.

Figure 2.2 shows an idealized case of aliasing. The original signal (solid line) was sampled at a rate slower than the Nyquist frequency. When the signal was reconstructed from the sampled points (dashed line), a much different signal resulted. The reconstructed signal did not correctly reflect the nature of the true signal. Aliasing is a common problem in wastewater control systems. As a rule, a signal (or value of interest) should be sampled at 5 to 20 times the time constant to obtain a good approximation of the original signal.

*P*ROCESS DYNAMICS

Process dynamics describes the change and rate of change in a process. Often, a set of simultaneous differential equations is used to describe the combination effect of transport by hydraulic flow and by gravity settling, and the effect of chemical change as a result of input and reaction velocity. From these equations a mathematical model may be prepared to represent the dynamics of a physical process. The digital computer has made it possi-

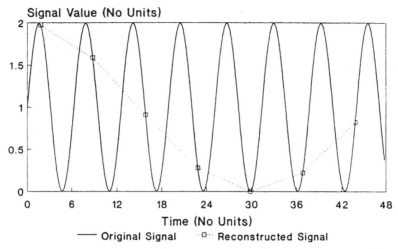

Figure 2.2   Effects of aliasing—original and reconstructed signals (WEF, 1993)

ble to use complex models to simulate processes and evaluate different control strategies.

Dynamic systems typically have time delays before the result of a process change can be seen. In addition, the processes are often nonlinear. Some examples of nonlinearity are as follows:

- The rate of clogging of a bar screen may be proportional to flow in normal weather, but in the event of high flow from a rain, material may be dislodged from the sewers and cause a greatly increased solids loading.
- The chlorination of an effluent containing ammonia is certainly nonlinear. At first the chlorine residual (mono- and dichloramine) increases with increasing chlorine dose. With further addition of chlorine, trichloramine is formed that can volatilize and/or decompose to nitrate and other products. The result of further chlorine addition is a decrease in the chlorine residual. Only after the breakpoint is reached does the residual again increase with added chlorine.
- In an activated-sludge final clarifier, when the solids input is greater than the solids removed in the underflow, the difference accumulates in the sludge blanket until the clarifier is full. Then, the difference between the inflow and underflow solids enters the effluent and causes a sudden increase in the effluent solids concentration.

Control schemes may be based on process dynamics and process models, but it is important to remember the purpose. Within the desired control

---

range, a linear approximation may be sufficient. The process model is adequate if the controller provides the desired results. Sometimes the desired result is beyond control of the chosen process variable (PV). Allowance for some deviation of the PV may be necessary to avoid overcorrection of a more important variable.

DEVELOPMENT OF TIME CONSTANTS

In the foregoing discussions, time is considered a variable. However, when the subject becomes control of a certain process in a specific vessel, then there are *time constants* in which time is the relationship between such parameters as distance and velocity or tank volume and flow. The rate of change in the process variable is related to the time constant of the process. When an analyzer that operates on periodic discrete samples is used, a delay called *dead time* occurs that includes the time of travel to the analyzer and the analysis time. This is additive to the process dead time.

The simplest, and often an adequate, model of a process is the first-order-plus-dead-time model. The dead time may result from imperfect (not instantaneous) mixing involving a stirred reactor, or it may include travel time to an analyzer. The first-order time constant for a simple stirred mixing tank at constant flow is volume divided by rate of flow, that is, the detention time. The process model and time constant are important in the design of a controller or in the tuning of a proportional-integral-derivative (PID) controller as described in chapters that follow.

While it is possible to calculate estimates of the time constants, they are normally determined as part of an on-line controller tuning procedure. Ziegler and Nichols (1942) described open-loop and closed-loop procedures to determine tuning constants. Åström and Hägglund (1983) described an improved closed-loop tuning procedure in which a relay control is used that gives reasonable control while data are logged to determine the tuning constants. These procedures are described in a more recent book by Åström and Hägglund (1988). Experiences with the Åström and Hägglund relay procedure in tuning chlorination controllers have been discussed by Garrett *et al.* (1993).

Olsson (Personal Communication, Lund Institute of Technology, 1990) has warned that the Ziegler–Nichols (Z–N) tuning constants tend to produce cycling of air supply in a dissolved oxygen control loop, and recommends using a smaller gain. The Z–N tuning constants were derived to give 1/4 decay ratio of setpoint overshoot; that is, each successive excursion (above or below) the setpoint decays to 1/4 the magnitude of the previous excursion. Perry and Chilton (1973) give a table of suggested controller settings for 1/4 decay ratio and for no overshoot. The recommended gain for no overshoot in a PID loop is 1/3 the value of the gain for 1/4 decay ratio. Thus, it may be necessary to reduce controller gain to have the controller give the desired results.

The controller, whether it is a PID controller or some other, is ultimately the tool for implementing automatic control of the selected control strategy.

REFERENCES

Åström, K.J., and Hägglund, T. (1983) Automatic Tuning of Simple Regulators for Phase and Amplitude Margins Specifications. *Proc. IFAC Workshop Adaptive Syst. Control and Signal Process.*, San Francisco, Calif., 271.

Åström, K.J., and Hägglund, T. (1988) *Automatic Tuning of PID Controllers*. Instrum. Soc. Am., Research Triangle Park, N.C.

Bibbero, R.J. (1977) *Microprocessors in Instruments and Control.* John Wiley & Sons, Inc., New York, N.Y., 19.

Garrett, Jr., M.T. (1990) *Wastewater Treatment Process Energy Optimization: A Study of the Control of Blowers and Other Loads at a Plant in Houston, Texas.* Public Technology, Inc., Washington, D.C.

Garrett, Jr., M.T., et al. (1993) Experience with the Relay Procedure for Tuning Controllers in Automatic Control of Chlorination. *Proc. 6th IAWQ Workshop*, Banff, Alberta; Hamilton, Ont., Can.

Directory of Instrumentation (1996). Inst. Soc. Am. Serv., Inc., Research Triangle Park, N.C.

Olsson, G., and Piani, G. (1992) *Computer Systems for Automation and Control.* Prentice Hall International, Ltd. (U.K.), Hemel Hempstead, Herts., 144.

Perry, R.H., and Chilton, C.H. (1973) *Chemical Engineers' Handbook.* 5th Ed., McGraw-Hill Book Co., Inc., New York, N.Y., 22.

Water Environment Federation (1993) *Instrumentation in Wastewater Treatment Facilities.* Manual of Practice No. 21, Alexandria, Va., 238.

Ziegler, J.G., and Nichols, N.B. (1942) Optimum Settings for Automatic Controllers. *Trans. ASME.*, **64**, 759.

Chapter 3
General Control Concepts

CONTROL THEORY

The most common use for instrumentation in wastewater treatment plants (WWTPs) is process monitoring or data logging. Instrumentation has the advantage of producing information describing the state of the system on a frequent, or even continuous, basis. This information can be used to show in more detail what occurred over a particular time period. It can also be used to enable the operator to respond more rapidly to changes in the system. If the operator action can be converted to electronic control, as in the case of pumps, blowers, and valves, and rules governing the response to process changes can be expressed formally, then automatic control becomes a logical choice. It also has the advantage of resulting in more timely actions than manual control. This can keep the process performance closer to the desired result and can do so at reduced cost.

The better the process behavior and the inputs to the process are understood, the more closely it is possible to control the results. Fortunately, it is possible to greatly improve process performance even if the process behavior is only known approximately. A process is "known" in terms of a mathematical model of its behavior. The task of determining the appropriate model to use and its parameters is called *process identification*. With the model in hand, a mathematical control law can be devised.

PROCESS IDENTIFICATION AND MODELING

A mathematical model of a process is a set of equations that can, to some degree of realism, predict the response of the process to any given set of input conditions, or forcings. A model by its nature is not perfect; it is essentially an analogy to the real process that can be tested and otherwise manipulated to predict what the real process might do. It is necessary to strike a balance between the simplicity of a model and its ability to predict the process behavior realistically.

A model may be *fundamental* or *empirical*. A fundamental model is derived mathematically from the laws of physics, chemistry, and biology that describe the parts of the process. Examples of these fundamental laws include mass, energy, or force balance laws, and laws of kinetics applied to chemical or biological reactions. Particular parameters of the fundamental processes will typically need to be measured by experiment to complete the model.

An empirical model attempts to describe the behavior of a process directly without considering the internal details of the process. The model is developed by introducing disturbances to the process inputs and observing the response. The disturbances may have the form of a step change in the input, a pulse forcing, or an impulse.

Mathematical forms for the empirical model are postulated based on the observed responses and then tested by further experiments until a satisfactory model is developed. Sometimes, a simple fundamental model may have the same mathematical form of an empirical model.

Most processes can be classified as being of one of the following types:

- Steady state,
- First-order lag, or
- Second-order lag.

In addition, a dead time, or transport lag, may occur with each of these models.

A *steady-state* process is one in which no relevant variables (such as flow or concentrations) are changing with time. Real processes may be approximated as steady state if the changes in inputs are slow relative to the dynamics of the process itself. Steady state is sometimes confused with *equilibrium*. An equilibrium process is one in which reactions occur so rapidly that for practical purposes they are instantaneous. This is the case, for example, in acid–base reactions in a flash-mixer. In this and all equilibrium processes, the output (pH) depends only on the inputs (amounts of acid and alkalinity) and will not depend appreciably on the time since any changes were made.

The model for a steady-state process will be an algebraic expression. A common form is the simple proportional relationship. The constant of proportionality is known as the process gain, KP. If the input to a process is X and the output is Y:

$$Y = KP \times X \tag{3.1}$$

As an example, the production of solids in an activated-sludge process might be found to be proportional to biochemical oxygen demand (BOD) loading for a certain range of operating conditions. The gain in this case would be the observed yield of the process (kilograms [pounds] of solids produced per kilogram [pound] of BOD removed).

Most processes have appreciable dynamics. When input conditions are changed, the outputs adjust gradually to the steady-state response. The mathematical models describing dynamic systems incorporate differential equations.

The simplest type of dynamic system is the first-order system, which is characterized by the presence of some form of capacity and resistance. An example is a completely mixed constant-volume tank with flow that is added to it at a variable rate and contains a variable concentration of some conservative (nonreactive) substance, such as common salt. The steady-state behavior of this system can also be described by a gain, which is the ratio of the change in output to changes in input. The gain would be 1 in this example. That is, a unit change in input concentration results in a unit change in output concentration. The presence of chemical reactions could result in other values for the gain.

The system approaches a new steady state rapidly at first, and then more and more slowly, in a type of response called *exponential decay*. The dynamics of a first-order process can be described by a single parameter in addition to the gain. This is called the time constant, τ_P, which has units of time. The higher the flow rate, the faster the output responds to changes in influent concentration. In the presence of a chemical reaction that consumes the salt, the dynamics would be slowed and τ_P would be increased. Numerically, τ_P is equal to the time required to achieve 62.3% of the difference between initial and final conditions. After three intervals of τ_P have passed, a first-order system will have changed approximately 95% towards steady state.

General Control Concepts   *17*

A rapidly changing process would have a short time constant. For example, a change in air flow rate to an aeration basin causes the dissolved oxygen (DO) level to adjust within several minutes. Other changes occur more slowly: in the activated-sludge process, the time constant for changes in biomass is the sludge age, which is measured in days. Typical time constants for wastewater processes are listed in Table 3.1.

Many systems exhibit more complex behavior than first-order systems. Systems that have two capacities and resistances, such as two first-order systems connected in series, are second-order processes. The response of such a system to a step change in input would resemble a sigmoidal curve. Such a system would be described by two time constants in addition to the process gain.

Another type of second-order process is one that exhibits both capacity and momentum, typically with resistance as well. The best known of this type is the spring-mass-dashpot (shock absorber) system. Any second-order system is capable of oscillations around the steady-state value. For example, when the fender of a car is pushed downward, the spring will return it to its original position. If the shock absorber is not providing sufficient resistance, the momentum of the car will cause it to overshoot the original position. The car will then vibrate up and down with decreasing amplitude until it settles to the steady-state position. This phenomenon is known as *damped oscillation* or, more specifically, as an *underdamped system*. If the shock absorber is providing enough resistance, the car will settle down without overshooting the mark. This response is called a *critically damped* system.

Another common factor in the dynamic response of processes is dead time, or transportation lag. The typical example is a long pipeline connected to a process. A 300-m (1 000-ft) long 0.15-m (6-in.) inner diameter pipe has a volume of almost 5 500 L (1 500 gal). If the flow through the

Table 3.1   **Typical time constants for wastewater processes (WEF, 1993)**

Process or component	Time constant
Blowers/surge	<1 Second
Header pressure and air-flow rate	A few seconds
Effluent suspended solids	Minutes
Hydraulic flow rate	5–10 Minutes
Dissolved oxygen concentration	10–30 Minutes
Aeration basins/mixed liquor suspended solids concentration	A few hours
Aerobic solids growth rate	Days
Anaerobic solids growth rate	Weeks
Influent temperature	Months

pipe were 6 L/s (100 gpm), the discharge from the pipe would be representative of what went into the pipe 15 minutes earlier. If this pipe were connected to some other process, such as a first-order process, then the response measured at the end of the pipe would be called *first-order plus dead time*. One of the reasons that dead time is significant is that it is difficult to account for when using the type of mathematical models typically applied to processes. As a result, the mathematical theories generally employ approximations to *pure dead time*. The *Smith Predictor* is an example of such an approximation.

Processes may be of higher than second order. Any process consisting of several processes in series would have one or more time constants for each subunit. In most practical situations these can be approximated by first- or second-order systems with dead time.

CONTROL STRATEGIES

Once process identification has provided us with sufficient understanding of the behavior of a process, we can devise methods to control it. First it is necessary to define some terminology and some basic concepts. The variables that describe a process may be classified as *input variables*, *process parameters*, and *output variables*.

Input variables include *disturbance variables*, which are not controlled by the operator, and *manipulative variables*, which can be controlled. For example, in the activated-sludge process, influent flow and BOD are typically not controllable, and thus are disturbance variables. Manipulative variables include return flow rate, air flow rate, and waste solids flow rate. Note that although waste flow rate represents an output with respect to a mass balance of the system, in the control scheme it is an input. Any type of change to an input, whether imposed or not, is termed a *forcing*.

Output variables include any variables that respond to changes in the inputs. This includes such things as the composition of streams leaving the process, as well as compositions within. In the activated-sludge process, effluent suspended solids and BOD are outputs, as are mixed liquor suspended solids in the aeration tank. Note that waste activated sludge suspended solids concentration is an output, even though its flow rate is an input.

Process parameters are those that describe the system behavior. These may be the empirical values such as gain and time constant described above, or more fundamental ones such as growth rate of microbes or the partition coefficient of an adsorption process.

The purpose of the proposed control scheme must be defined in a precise way in terms of measurable output variables. The resulting definition is known as the *control objective* or *objective function*. The objective may be as simple as "minimize effluent BOD" or "maintain dissolved oxygen level close to 2.0 mg/L," or it may be more complex, as in "minimize solids disposal and energy costs while keeping BOD below effluent permit requirements." Note that some likely objectives, such as cost, may conflict with other objectives, such as effluent quality. In cases such as these, it is generally possible to assign weights to each individual objective or express some as a constraint, as in permit requirements.

In general, the control function will have the form of a minimization. If the control objective is to maintain some output variable close to a desired value, the desired value is termed the *setpoint*, *SP*. The objective function to be minimized in this case is typically the error, *E*. This is the actual value of the variable (the process variable, *PV*) minus the setpoint.

$$E = PV - SP \tag{3.2}$$

In the example of DO control, if the setpoint is 2.0 mg/L and the actual DO is 1.2 mg/L, the error is equal to 0.8 mg/L.

It is essential that the outputs used in the objective function actually be related to the inputs by the process model. This is termed the *controllability problem* and may be of special concern in biological treatment processes. In the activated-sludge example, there is no fundamental model to predict effluent suspended solids on the basis of inputs such as waste rate, food-to-microorganism ratio, or amount of biomass. The control approach typically selected by operators, therefore, is an indirect one. A surrogate output parameter that can be controlled, such as sludge age, is used instead of the uncontrollable effluent suspended solids. This selection involves an assumption that the desired objective is functionally related to the surrogate parameter. The actual relationship in this case is generally unknown, but operators will select the setpoint sludge age based on trial-and-error experience and experimentation.

Another complication common to biological processes is that the process parameters may change because of measured and unmeasured changes in the inputs. The input composition is typically measured by gross parameters such as BOD or chemical oxygen demand. However, biomass may respond differently to wastewaters having the same BOD but different chemical compositions. Also, a natural succession of microbial organisms could change the system response even if inputs do not change. Wastewater treatment plant operators know this intuitively, and they respond to unexpected changes in plant performance by continuing their trial-and-error search for optimum conditions.

INTEGRATED CONTROL STRATEGIES

Control strategies in industrial processes are typically based on a feedback loop—system output is monitored and control actions are based on the deviation from the desired system state. Control can be classified according to the automation as follows:

- Manual control—operators observe the process and implement control actions manually.
- Local automatic control—individual control loops are automated, and the control elements are modulated based on the measurements.
- Supervisory systemwide control—a comprehensive set of measurements is presented to the operators using a supervisory control and system data acquisition (SCADA) system, typically to a central control room. The decisions are still made by the operators, but the control actions can be issued remotely from the central console.
- Integrated systemwide automatic control—signals from the field measurements are processed by computer programs, and the control actions are determined by algorithms.

An integrated control strategy connects all process control loops within the WWTP into a comprehensive and coordinated system that manipulates the operation of all the units during any process conditions. The task of the integrated control strategy is to assess the state of the process on line and determine the appropriate control action.

An integrated control strategy serves as an ''umbrella'' for system operation. Depending on the state of the system, the objective of the overall control strategy can be as follows:

- Prevent gross process failure;
- Archive performance necessary for meeting permit levels; and
- Optimize performance in terms of cost (for example, energy and chemical costs).

The goals of the control strategy, as shown above, have clearly been listed in the order of priority. The overall control strategy receives information about the process state from the sensors, typically via SCADA. The appropriate control decision is made by the algorithms and passed on to the local controllers.

CONTROLLERS

Controllers are devices that take signals from sensors on process inputs and outputs and produce a change in one or more manipulated variables by means of some type of actuator. The controllers implement a control strategy that, although designed in mathematical terms, is realized by mechanical or electronic means.

A familiar example is the home thermostat, which senses the temperature of a room by the bending of a bimetallic strip and actuates a heating system by using the movement to open or close a circuit. Until recently, more sophisticated control strategies were implemented in the chemical process industry using pneumatic devices operating on air pressure. Pneumatic devices were designed to respond analogously to the types of mathematical relationships used in controllers, such as adding, subtracting, or multiplying signals (in terms of air pressure), or even operations such as integration or finding derivatives. Digital electronic elements have become more important in recent years, performing these functions by computer.

Control strategies may be classified as *feedback*, *feedforward*, or a combination of the two. Feedback controllers perform their manipulation based only on process outputs, whereas feedforward controllers are based only on process inputs. Only single-input–single-output controllers will be discussed here. Controllers may have multiple inputs or outputs. Examples include cascade controllers, override control, and split-range control (Stephanopoulous, 1984).

FEEDBACK CONTROL

Feedback control involves the use of measurements of a process output variable to make adjustments to an input variable, which becomes the manipulated variable. Examples include

- Adjusting pump speed to control wet-well level,
- Adjusting lime dosage to control pH,
- Controlling DO concentration by blower speed or valve position,
- Controlling sludge age by manipulating waste flow rate, or
- Controlling temperature by turning a heater on and off.

Note that in many situations, the human operator provides the feedback by making adjustments manually based on periodic observations of process performance. This discussion will focus on automatic methods of control. Systems that are left to run by themselves without any systematic form of feedback are termed *open-loop* systems. When a controller is added they are called *closed-loop* systems.

The simplest form of feedback control is on–off control. The obvious example of this is the ordinary home thermostat. The thermostat responds to the temperature dropping below the setpoint by turning the furnace on. When the temperature has risen sufficiently, it is then turned off. A lag effect built into the system is necessary to prevent the on–off cycles from becoming too rapid. Another example is control of water level using a float switch to control a constant-speed pump. A problem with this type of system is that the controlled variable, in these cases temperature or water level, must oscillate around the setpoint. A particular advantage with energy-consuming devices is that they could be run only at their most efficient level. Smoother response would result if the device could be throttled, but this often reduces energy efficiency.

PROPORTIONAL-INTEGRAL-DERIVATIVE CONTROL. Other than the on–off controller, the most popular type of control in industry is the proportional-integral-derivative, or PID, controller. The PID controller can reduce or eliminate the oscillations found with on–off control. Each of the three parts of the PID controller can be considered separately. A useful example to help understand the PID controller is to apply it to control of an automobile's speed, such as might be done by the cruise control option available in many cars.

The basic action of this type of controller is proportional control (P control). Consider a car on a flat road and maintaining a speed of 88 km/h (55 mph). If the car slows because of a change in wind or slope of the road, the driver depresses the accelerator. The proportional controller mimics this by depressing the accelerator in proportion to the error, E (the actual speed, PV, minus the desired speed, SP). The accelerator position is the manipulated variable, M. Mathematically, this type of control can be expressed as

$$M = M' + K_C \times E \qquad (3.3)$$

M' is the setting of the manipulated variable when the error is zero. K_C is called the *proportional gain* of the controller and determines how far the accelerator is depressed for each increment of error. For example, its value might be 5 mm (0.25 in.) per kilometre (mile) per hour below the setpoint speed.

The response of the proportional controller depends on the value of the gain. If K_C were too large, a drop in speed would result in the accelerator being depressed too far, causing the speed to overshoot the setpoint, whereupon the reverse would happen, slowing the car too much. Thus, the value of K_C is related to the "stability" of the system. If it exceeds a critical value, a disturbance may send the system into oscillations. Stability is a crucial consideration in the design of all control systems.

Conversely, if K_C were too small, then a drop in speed would not be sufficiently compensated for, and the car's speed would settle at a new value with a nonzero error. This steady-state error is called the *controller offset*. It is a characteristic of proportional-only control that the offset cannot be completely eliminated while maintaining a stable system.

The addition of *integral* action can eliminate controller offset. Integral action adjusts the manipulated variable continuously at a rate proportional to the amount of error. If any offset is present, the integral action will continue to make changes until it is eliminated. Consider the example of the car with cruise control that encounters a hill. The proportional action depresses the accelerator, which provides partial compensation but results in an offset. Integral action continues to depress the accelerator until the setpoint speed is reached.

Mathematically, the combination of proportional and integral control (denoted PI control) is described as follows:

$$M = M' + K_C \times \left(E + \frac{1}{\tau_I} \int E dt\right) \tag{3.4}$$

Where τ_I is known as the *integral time constant*, or *reset time*. Its inverse, $1/\tau_I$, is referred to as *minutes per reset*, and determines how fast the controller increases its action in proportion to the amount of error. In cruise control, if the number of minutes per reset were large, the controller would depress the accelerator rapidly whenever the speed dropped. In this case the car would still be accelerating when the setpoint was reached, and the car would overshoot the desired speed. During overshoot, the controller would tend to act to correct itself, causing the system to oscillate around the setpoint. Depending on the values of the proportional gain and the integral reset time, the oscillations will either decrease until the system settles at the setpoint, or they may increase, indicating that the system is unstable.

The third type of action a PID controller can take is *derivative* action (D control). This action adjusts the manipulated variable in proportion to the rate of change of the process variable. In cruise control, if the speed changes, the accelerator position is moved in proportion to how fast the speed is changing. This helps dampen changes from large disturbances.

The equation for combined PID control is

$$M = M' + K_C \times \left(E + \frac{1}{\tau_I} \int E dt + \tau_D \frac{dE}{dt}\right) \tag{3.5}$$

Where τ_D is the derivative rate parameter that determines how much the controller responds in relation to the rate of change of error. Derivative action should not be used with processes with significant dead times because the derivative itself may add to the noise.

PROPORTIONAL-INTEGRAL-DERIVATIVE CONTROLLER TUNING. Selection of appropriate gain, reset, and rate parameters may be difficult. If the process can be represented by a simple analytical model, then the parameters can often be computed. Alternatively, the system can be tuned experimentally by introducing some type of disturbance to the system, observing the dynamic response, and calibrating a simplified model to that response. The tuning parameters can then be related to the measured dynamic response.

To compare controller settings, some measures of controller performance are needed. Examples include *minimum offset, one-quarter decay ratio*, or *minimum integral square error* (see Stephanopoulous, 1984).

Several tuning methods will be described briefly here:

- Rules of thumb,
- On-line trial-and-error,
- Cohen–Coon method,
- Ziegler–Nichols method,
- Analytical, and
- Computer simulation.

The first method is based on previous experience with certain types of control loops, such as control of flow, level, pressure, or temperature, which are common processes. For example, PI control is typically used for flow control (Luyben, 1973). The proportional gain is typically set low to reduce the effects of the noise that is inherent in many flow meters. The integral reset time is set low, corresponding to a large integral action, to achieve fast response to changes in the setpoint error.

The next three methods are experimental. They involve introducing deliberate disturbances to the system. Rules are used to compute the tuning constants from the way the system responds to the disturbance. Both the Cohen–Coon method and the Ziegler–Nichols method have some practical shortcomings. They are sometimes inaccurate and must be followed up by trial-and-error refinement. In many situations, the experiments that are required may not be feasible. For example, it may not be possible to achieve an initial steady state for the Cohen–Coon method. In these cases, the analytical or the computer simulation method could be used.

These last two methods require the existence of a fairly reliable mathematical model of the process. The analytical techniques use mathematical procedures called *Laplace-domain synthesis* and *frequency-domain synthesis* to compute values of the tuning parameters. If the models are too complex, even these methods may not be usable. In this case, the model can be used to produce a computer program to simulate the process and the PID controller. The tuning parameters can then be found by a combination of experimental and trial-and-error methods performed using the simulation in place of the real process.

Feedback control is sometimes used to control DO levels in activated-sludge WWTPs. The use of PI control has been reported for this purpose (Corder and Lee, 1986). A PID controller has been developed to control sludge age in an activated-sludge process by manipulation of the waste flow rate (Vaccari *et al.*, 1988).

*F*EEDFORWARD CONTROL

A feedback controller theoretically cannot achieve perfect control for the simple reason that it can only respond when an error is detected in the output. A way to improve on this situation is to detect disturbances at the input, before the system responds, and adjust the manipulated variable to cancel out the effect of the disturbance. Thus, the essential difference between feedback and feedforward control is that the former acts by compensating for changes, whereas the latter anticipates them.

An example found in many water and wastewater treatment plants is the control of chlorine residual. The chlorine dosage required depends on the chlorine demand of the water or wastewater and its flow rate. The residual could be controlled by a feedback controller based on a final chlorine concentration sensor. The sensor should take its sample from a point having an adequate detention time following chlorine dosage, so that sufficient time, approximately 15 minutes, will have passed for the chlorine demand to express itself. If the flow changes significantly, insufficient chlorine will be added to the stream for at least 15 minutes, until the sensor detects the change. In other words, there is a significant dead time in the system response. As described above, this can result in instability. Some feel that if proper mixing is provided, chlorine can be sampled almost immediately (1 to 2 minutes) after being fed.

If the chlorine demand is known to be fairly constant, an alternative control scheme would be to adjust the chlorine dosage in proportion to the water flow. This is a simple form of feedforward control known as *ratio control*. In ratio control, the ratio between a manipulated variable and an input is maintained constant by adjusting the manipulated variable. In this example, the ratio is that of chlorine dosage to water flow rate.

Ratio control can be implemented using a PID controller with integral and derivative action turned off and the setpoint set to zero. In this case, the output is proportional to the input because the error is equal to the input.

Ratio control ignores the dynamics of the process. In the chlorination example, changes in the water flow also change the detention time of the chlorine contact chamber. Thus, even though the chlorine demand and chlorine dosage are kept constant, the concentration being discharged might change with flow. If, however, the reactions occur quickly so that near equilibrium is achieved, then the dynamics are not important. When they are important, a dynamic feedforward controller can be used.

Dynamic feedforward control requires a model of the dynamic response of the system. The mathematical form of the controller will depend on the model for the system and cannot be generalized as simply as the PID controller for feedback control. The basic idea of dynamic feedforward control is to measure the input variables and the state of the system and to compute the value of the manipulated variable that would result in the system staying at the setpoint.

In principle, except for cases as described below, feedforward control is capable of maintaining the setpoint exactly, in spite of the presence of input disturbances. However, feedforward control is not used as frequently as feedback for the following reasons. First, it is not always easy to sense the disturbances adequately. An example is the difficulty of measuring BOD input to a biological treatment process. The second reason is that a suitable model, or model parameters, of a process is not always easy to determine, especially for a biological process in which the process parameters may change.

The third reason, and the exception mentioned above, is that the dynamics of many processes are such that they respond more slowly to changes in the manipulated variable than to changes in the inputs. When this occurs, the feedforward controller is said to be *physically unrealizable*. An example is the activated-sludge process. The sludge age of the unsteady-state process may change rapidly with changes in influent BOD loading, whereas the system responds more slowly to changes in waste solids flow rate. In such a situation, exact control is impossible, though partial control may still be achieved.

Another example of feedforward control is a design proposed to control DO in an activated-sludge WWTP (Corder and Lee, 1986). In this controller, the input to the controller is the BOD loading rate. Based on simulation experiments, it is predicted that the feedforward controller would reduce air flow rate by 20%.

COMBINED FEEDBACK/ FEEDFORWARD CONTROL

When feedforward control cannot be accurately or completely achieved, it may be combined with feedback control. The feedforward controller then reduces the amount of variation that must be handled by the feedback controller, improving its stability characteristics. The feedback controller reduces the perturbations that escape feedforward control.

The chlorination system in use in most water and wastewater treatment plants is an example of combined feedback/feedforward control. The feedforward loop is implemented as ratio control based on water flow rate as described above, and the feedback loop uses a residual chlorine analyzer to make further adjustments to the chlorine dosage. The ratio control can act fairly quickly and prevents flow-induced disturbances from affecting the chlorine residual. The concentration feedback loop compensates for residual dynamic effects of flow and for changes in chlorine demand of the water or wastewater. Because of the dead time in the feedback loop, its response is much slower than the feedforward component of the controller.

A second example of combined feedback/feedforward control found in some WWTPs is the system used to control methanol dosage added to a denitrification process (U.S. EPA, 1993). This algorithm is similar to the chlorination controller in that the feedforward component is ratio control based on flow, and the feedback variable is concentration. It is different from chlorination in that the concentration that is fed back is not that of the dosed compound, methanol, but rather is the effluent nitrate concentration. Thus, the feedback action compensates for the actual process performance.

ADVANCED CONTROLLERS

The availability of digital computers makes more sophisticated types of control feasible. These control methods include *direct digital control* and *adaptive control*.

Direct digital control can be implemented in the same way as the analog (nondigital) controllers described above. A characteristic of digital control is that measurements are made at discrete time intervals instead of continuously. If the intervals are small, digital control can simply mimic analog. However, the computational capability inherent in digital control makes other types of controllers possible.

In adaptive control the tuning constants may change with time in response to changes in the process. Self-tuning regulators are an example of this. Essentially, the controller monitors the response of the system to disturbances and controller forcings and modifies the tuning parameters as

---

needed. This type of control is useful when the process model is not known beforehand or may change significantly with time. Biological wastewater treatment processes are a good example of the latter.

*R*EFERENCES

Corder, G.D., and Lee, P.L. (1986) Feedforward Control of a Wastewater Plant. *Water Res.* (G.B.), **20**, 301.

Luyben, W.L. (1973) *Process Modeling, Simulation, and Control for Chemical Engineers.* McGraw-Hill, Inc., New York, N.Y.

Stephanopoulous, G. (1984) *Chemical Process Control, An Introduction to Theory and Practice.* Prentice-Hall, Inc., Englewood Cliffs, N.J.

U.S. Environmental Protection Agency (1993) *Nitrogen Control.* EPA-625/R-93/010, Office Res. Dev., Office of Water, Washington, D.C.

Vaccari, D.A., *et al.* (1988) Feedback Control of Activated Sludge Waste Rate. *J. Water Pollut. Control Fed.*, **60**, 1979.

Water Environment Federation (1993) *Instrumentation in Wastewater Treatment Facilities.* Manual of Practice No. 21, Alexandria, Va.

Chapter 4
Expressing Process Control Concepts

PROCESS CONTROL DOCUMENTATION

The documentation of automated process control concepts is an important step in the automation procedure. The documentation provides a hard copy of the process control ideas of the engineer and a common medium that interested parties can use to communicate effectively and efficiently.

There are many ways to document process control concepts. Several methods will be discussed within this chapter:

- Control narratives,
- Pseudolanguage programs,
- Relay ladder logic,
- Logic diagrams,

- Flow charts, and
- Process and instrumentation diagrams.

Whatever method of documentation is used, there are essential pieces of information that should be included in the documentation package:

- Title, identification, and brief description of the operational unit process to be controlled;
- Brief description of the physical system, such as wet well volume and dimensions, pump capacity–head relationship, pipe diameter, and temperature and pressure data;
- Information relating to the process that is external to the process control that will be documented, such as local controls or nonautomated controls;
- Input and output signal lists, including signal names and/or numbers;
- Logic and equations required for process control;
- Logic and equations required for the operator interface, such as alarm messages, process graphic displays, and periodic reporting needs;
- Database constants, such as timing values, dimensions, conversion factors, and other constants that may enter the equations and logic; and
- Physical constraints of the system, such as wet well volume, volume per foot, pump capacity–head relationship.

There are many ways to express process control concepts. This chapter will present several accepted methods of expressing process control strategies.

For ease in following each method, the example for each method describes the same sample process control scheme—that is, the control of a bar screen. To maintain simplicity, the example uses only the on–off control portion of the equipment; no alarms or shutdown logic has been included. Additional controls such as permissives, local and remote signaling, and interlocks to allow or prohibit operation have not been included in the example, although these items may be desirable for actual schemes.

CONTROL NARRATIVES

Text descriptions of process control strategies are the simplest method of documentation. Written in plain English, they are easy to understand. Because no technical skill or training is required to read them, control narratives can be read and understood by a wider audience than the other documentation methods. Higher management, who may not have the experience or training to comprehend other methods, can read and understand a text description easily. Similarly, a wastewater treatment plant (WWTP) operator or technician can also read a text description without training in any of the graphics-based methods (see the example in Figure 4.1).

The ease in writing the control narrative is a major advantage, but it is also a disadvantage. Because there is no format or order inherent in the method, it requires extra care in development to ensure that rigorous thought has been applied in developing the description.

PSEUDOLANGUAGE PROGRAMS

This method also uses text descriptions but introduces a structure that may be helpful. The structure helps the writer by organizing the material into a logically formatted description. It is called a *pseudolanguage* because it is not any specific programming language, such as Basic or Fortran, nor any specific manufacturer's proprietary language. This generality may be helpful if the control documentation is prepared before the selection of the vendor. The documentation would then be general, so as not to limit the number of vendors from which a supplier will be chosen.

Because it is not a true language, there are no rules for syntax or format. However, because its structure is an advantage, the presentation should be consistent as to format, syntax, and content (see the example in Figure 4.2).

The bar screen is controlled by a HAND-OFF-AUTO selector switch located adjacent to the screen.

When the HAND-OFF-AUTO selector switch is in the HAND position, the screen will operate continuously until the switch is moved to the OFF position.

When the HAND-OFF-AUTO selector switch is in the OFF position, the screen will not operate.

When the HAND-OFF-AUTO selector switch is in the AUTO position, the screen will operate periodically. Two timing relays will control the duration of screen operation and the interval between screen operations. Differential level across the screen is monitored. If high differential level is detected, and if the HAND-OFF-AUTO selector switch is in the AUTO mode and the screen is not running, the screen will immediately start its cleaning operation and operate for the duration of the cleaning cycle.

Figure 4.1   Control narrative

```
50     IF Screen is in AUTO mode
       THEN Start Interval Timer; go to line 100
       ELSE exit

100    IF a high differential level exists
       THEN go to line 200
       ELSE IF Interval Timer has expired
               THEN go to line 200
               ELSE go to line 100

200    Reset Interval Timer; start Duration Timer

250    IF Screen is in AUTO mode
       THEN Run Screen
       ELSE go to line 300
       IF Duration Timer has expired
       THEN IF a high differential level exists
               THEN Reset Duration Timer; go to line 200
               ELSE go to line 300
       ELSE go to line 250

300    Stop Screen; Reset Duration Timer; go to line 50
```

Figure 4.2   Pseudolanguage program

LADDER LOGIC

Relay ladder logic is an accepted form for expressing control concepts. Ladder diagrams, also called *elementary* or *schematic* diagrams, were used as wiring and control diagrams for many years when binary logic control was based exclusively on relays and timers. The diagrams have been adapted to document software logic for programmable logic controllers (PLCs) or computers. The example herein (Figure 4.3) uses symbols from the National Electrical Manufacturers Association Standard ICS-1 (NEMA, 1993). For microprocessor-based equipment, the symbols used may be simpler to allow printing using a nongraphics-type printer. Because ladder logic diagrams are wiring diagrams, they are often well understood by technicians and electricians familiar with relay ladder logic, which can be an advantage during start-up or trouble-shooting.

UNDERSTANDING LADDER DIAGRAMS. This section is written using the example shown in Figure 4.3, which is a diagram used to show control of a motor using actual relays and timers. Diagrams generated by PLC or computer software often employ some simplifications to allow for nongraphics printers (such as depicting a relay as two adjacent parentheses instead of a circle), or to speed the printing process (such as not drawing the right-hand rail).

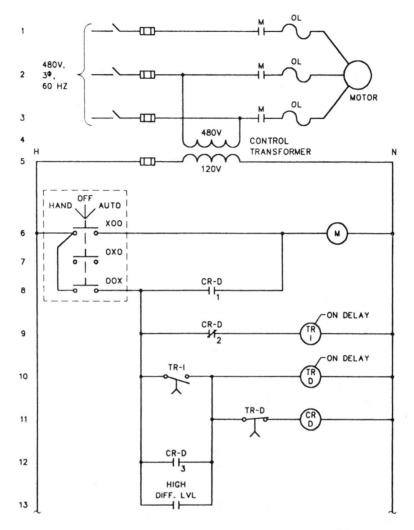

Figure 4.3 **Ladder logic diagram (CR = control relay; M = motor starter; TR = time delay relay; and OL = overload)**

Rails. The two outermost vertical lines are the rails of the ladder. In the example, the left-hand rail is hot (120 V-ac) and the right-hand rail is neutral. Current flows when a completed path exists between the two rails.

The horizontal lines between the rails are the rungs of the ladder. The example includes numbers for each rung to assist in identification. The numbers are to the left of the left-hand rail.

Relay. The relay is, as one might guess, the key to relay ladder logic. Although a description of what a relay is, and why it operates, is beyond the scope of this chapter, its operation is simple. Relays have two major parts: the coil and its associated contacts. When the relay coil (shown as a circle on the example) is energized, the relay contacts change state: open contacts (shown as two short vertical lines) close and allow current flow, and closed contacts (shown as two short vertical lines with a diagonal line crossing them) open and inhibit current flow.

The example shows three types of relays. Each type is described below:

- *Control relay*: labeled CR-D in the Figure 4.3, this is a standard type relay. There are three contacts associated with relay CR-D. Each contact is numbered for identification. Contacts CR-D1 and CR-D3 are shown open, and CR-D2 is shown closed. By convention, relay contacts are shown in the de-energized state, that is, the state they assume when the relay is not energized.
- *Motor starter*: labeled M in the example, this type has high-current-capacity contacts to carry the 480-V power to drive the motor.
- *Time delay relay*: labeled TR-I and TR-D in the example, this type introduces a delay between the time the coil is energized (or de-energized) and the contacts change state. In the example, both time delay relays are on-delay timers, in which a contact state change is delayed for a period of time after the coil is energized. When the coil is de-energized, the contacts revert immediately to their previous state. Contact TR-I is shown open; it will close when the timing period expires. Contact TR-D is shown closed; it will open when the timing period expires.

Selector Switch. The depiction in the example of a selector switch shows a three-position selector switch. Three T-shaped symbols show contacts that close depending on the mode selected. The notation "XOO" indicates that a particular contact is closed when in the first (HAND) mode; "OOX" indicates that the contact is closed in the third (AUTO) mode.

READING THE EXAMPLE. When the selector switch is in the HAND position (ladder rung 6), current will flow through the motor (M) coil, energizing it, which in turn energizes the motor (see ladder rungs 1, 2, and 3). The M coil will remain energized until the selector switch is moved from the HAND position.

When the selector switch is in the OFF position (ladder rung 7), there is no path to deliver power to the M coil, and the screen cannot start. The screen will stop if it was running when the switch was moved to the OFF position.

When the selector switch is in the AUTO position (ladder rung 8), current will flow through the M coil only if relay contact CR-D1 is closed. The remaining rungs of the ladder (9 through 13) include logic to operate

CR-D1, and this logic only operates in the AUTO mode. Operation is as follows:

- Rung 9: when the selector switch is placed in AUTO mode, relay contact CR-D2 is already closed, and current flows through timing relay TR-I. TR-I starts timing the interval period.
- Rung 10: when TR-I completes its interval timing, the TR-I contact on rung 10 closes, energizing timing relay TR-D (rung 10) and, because the TR-D contact is already closed, also energizing relay CR-D (rung 11). As stated previously, energizing CR-D closes the CR-D1 relay contact (rung 8), which in turn energizes the M coil, starting the motor. Energizing CR-D also opens CR-D2, resetting timing relay TR-I.
- Rung 11: when the timing relay TR-D completes its duration timing, the TR-D contact opens, de-energizing CR-D, which will de-energize the M coil and the motor. De-energizing CR-D will close CR-D2, restarting interval timer TR-I.
- Rung 12: when relay CR-D is energized by the TR-I contact, notice that on rung 9, contact CR-D2 opens, thereby removing power from relay TR-I. This, in turn, will cause contact TR-I to immediately reopen, removing power from both the TR-D and CR-D relays and stopping the motor immediately after starting. The relay contact CR-D3 provides another path, parallel to the open contact TR-I, which permits power to remain to TR-D and CR-D.
- Rung 13: a high-differential-level contact, closed when the differential reaches a high level, will immediately energize the duration timer (TR-D) and control relay CR-D, which starts the motor.

Logic diagrams

Logic diagrams show process logic by means of Boolean operators. These diagrams are, at their simplest, collections of AND and OR gates. In the example, timing modules must be included as well. In addition, several memory modules are used to hold an output until reset. These symbols are explained in detail in Instrument Society of America (ISA) Standard S-5.2 (ISA, 1976).

UNDERSTANDING LOGIC DIAGRAMS. This section is written using the example shown in Figure 4.4. Each module has its inputs on the left and its output on the right; the large rectangles on the extreme left show input signals, and the large rectangle on the extreme right shows a signal output. Where multiple modules of the same type are used in the example,

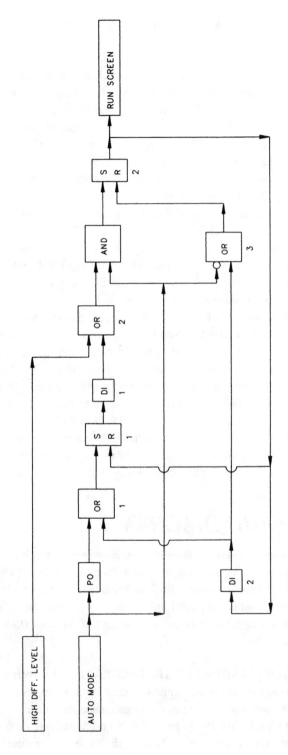

Figure 4.4 Logic diagram (DI = delay initiation module; PO = pulse output module; S = setpoint; and R = reset)

they are identified by a number adjacent to the module. Several types of gates and modules are used in the example. A brief description follows:

- *OR gate*: the output of an OR gate is true if any input to the gate is true.
- *AND gate*: the output of an AND gate is true if all the inputs to the gate are true.
- *Delay initiation (DI) module*: when the input to a DI module turns true, the module's timer starts timing (the module output remains false). As long as the input remains true, the timer continues to run. When the timer expires, the output turns true and remains true until the input turns false.
- *Pulse output (PO) module*: when the input to a PO module turns true, the output turns true for a set time and then turns false.
- *Memory module*: there are two inputs to a memory module, called the S (for set) input and the R (for reset) input. When the S input is true, the module output turns true and will remain true regardless of the subsequent state of the S input. The output will turn false when the R input turns true, and the output will remain false regardless of the subsequent state of the R input.
- *Inversion*: the small circle on the input side of a module inverts the sense of the signal. If the signal is false, the module will respond as if the signal were true, and vice versa.

READING THE EXAMPLE. When the AUTO MODE signal turns true, the PO module output turns true. OR gate 1 turns true, and memory module 1 turns true. DI module 1 starts timing (DI module 1 times the interval during which the screen does not run). When DI module 1 times out, its output turns true, turning the OR gate 2 output true. If the AUTO MODE signal is still true, the AND gate output turns true, and the memory module 2 output turns true. This starts the screen running and starts DI module 2 timing (DI module 2 times the duration during which the screen runs). In addition, the output of memory module 2 resets memory module 1, turning the output of DI module 1 false.

When DI module 2 times out, its output turns true, which will turn the output of OR gate 3 true, which in turn resets memory module 2. This stops the screen and turns the output of DI module 2 false. When the output of DI module 2 turns true, its output will also turn the OR gate 1 output true, setting memory module 1 output true, starting the timer of DI module 1 again.

A high-differential-level signal will immediately turn the output of OR gate 2 true, which starts the screen if it is in the automatic mode.

If the AUTO MODE signal goes false, it will turn the OR gate 3 output true, resetting memory module 2 and turning the screen off.

Flow Charts

This method of documentation is easy to read and to write. The example uses mainly decision blocks, whereby the logic "flow" diverges depending on whether a question is answered "yes" or "no." The text within the blocks can be written conversationally as used in the example, making the chart easy to read, or it can be written using Boolean symbols and signal numbers, making the programming easier.

UNDERSTANDING FLOW CHARTS. This section is written using the example shown in Figure 4.5. Logic flow progresses as shown by the arrows connecting the logic symbols. Several types of blocks are used in the example. A brief description follows.

- *Decision blocks*: shown as diamonds in the example, these blocks "decide" which logic path to take, dependent on the logic shown within the block. If the answer to the question within the decision block is "yes," the logic proceeds downward. If the answer to the question is "no," the logic proceeds rightward. Multiple decision blocks in the example are numbered for identification.
- *Process blocks*: shown as rectangles in the example, these blocks indicate actions the program takes that are internal to the program.
- *Input/output blocks*: shown as parallelograms in the example, the blocks indicate actions the program takes that are external to the program, such as an output to the field.

READING THE EXAMPLE. After the program starts, the first decision block is encountered. If the screen is in the automatic mode, the program will continue. If the screen is not in the automatic mode, the program will end. When the screen is in the automatic mode, the program will start the interval timer and continue.

The second decision block is next. If a high differential level exists, the program will reset the interval timer, start the duration timer, and advance to the fourth decision block. If a high differential level does not exist, the program will encounter the third decision block. If the interval timer has expired, the program will rest the interval timer, start the duration timer, and advance to the fourth decision block. If the interval timer has not expired, the program will wait and then return to the second decision block.

The fourth decision block is next. If the screen is still in the automatic mode, the logic will start the screen and continue. If the screen is not in the automatic mode, the logic will stop the screen, reset the duration timer, and go to the first decision block.

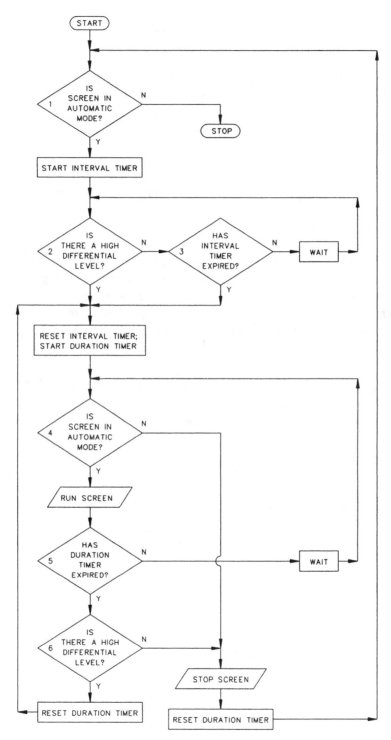

Figure 4.5   Flow chart

The fifth decision block is next. If the duration timer has expired, the program will continue. If the duration timer has not expired, the program will wait and then return to the fourth decision block.

The sixth decision block is next. If a high differential level exists, the logic will reset the duration timer, reset the interval timer, start the duration timer, and return to the fourth decision block. If a high differential level does not exist, the logic will stop the screen, reset the duration timer, and return to the first decision block.

PROCESS AND INSTRUMENTATION DIAGRAMS

Process and instrumentation diagrams are defined in ISA Standard S-5.1 (ISA, 1984). They are widely used. As shown in the example in Figure 4.6, the diagram shows the components and their wiring in a general way but does not detail the interlock or the operational logic. Process and instrumentation diagrams are complementary to the other forms of process control

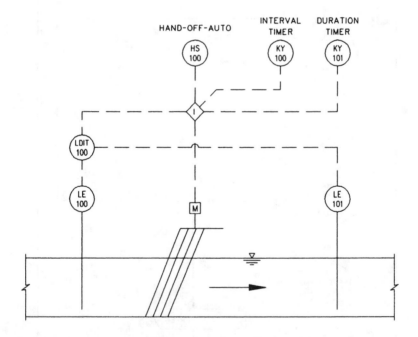

Figure 4.6   Process and instrumentation diagram (LE = level element; LDIT = differential level indicating transmitter; HS = HAND-OFF-AUTO switch; and KY = timers)

logic documentation. Process and instrumentation diagrams show the physical devices and interconnections among devices.

UNDERSTANDING PROCESS AND INSTRUMENTATION DIAGRAMS. This section is written using the example shown in Figure 4.6, that is, a diagram used to show control of a motor using discrete instruments. The screen and channel are shown at the bottom of the diagram; each instrument is shown as a circle. Symbols other than circles are used for logic within computer systems or PLCs. Connections between instruments are shown as dashed lines. The diamond indicates an interlock or detailed logic. This logic would be explained in detail, perhaps using one of the other methods of expressing process control concepts mentioned previously. Several types of instrument circles are shown on the example. A brief explanation follows:

- *Field-mounted instruments*: these are shown as plain circles (no line within the circle). The instrument tag number is shown inside the circle. The letters used for the tag number in the example are based on ISA Standard S-5.1 (ISA, 1984).
- *Primary location instruments*: these are shown as a circle bisected by a line. The primary location is typically a control panel. If the instrument is normally accessible to the operator (for example, front panel mounted), the bisecting line is continuous. If the instrument is not normally accessible to the operator (for example, internal to the panel), the bisecting line is dashed.

READING THE EXAMPLE. The example shows two field-mounted level elements (LE), one upstream and one downstream of the screen. The level elements are connected to a field-mounted differential level indicating transmitter (LDIT). The HAND-OFF-AUTO hand switch (HS) is mounted on the front of a control panel, and the two timers (KY) are mounted within the control panel.

*R*EFERENCES

Instrumentation Society of America (1984) *Instrumentation Symbols and Identification*. ISA Standard S-5.1, Research Triangle Park, N.C.
Instrumentation Society of America (1976) *Binary Logic Diagrams for Process Operations*. ISA Standard S-5.2, Research Triangle Park, N.C.
National Electrical Manufacturers Association (1993) *General Standards for Industrial Control and Systems*. Standard ICS-1, Washington, D.C.

Chapter 5
Lift Stations

CONTROL OBJECTIVES

The primary control objective of lift stations is to "lift" the entire incoming wastewater flow to the receiving unit process without overflowing the wet well. If required by the process, another primary control objective is to dampen variations in pumped flow rate.

Variable flow pumping does not always provide sufficient damping of discharge flow rate. Where appropriate, wastewater treatment plant (WWTP) design includes equalization basins of large storage capacity. Basins also serve as large wet wells, which aid in damping discharge flow rate.

Secondary control objectives must be met as well. These include protecting station electrical and mechanical equipment, rotating pumps to equalize wear, alarming abnormal conditions, and preventing overflows in both upstream and downstream unit processes. Another secondary control objective is matching the lift station's control strategy with the overall design philosophy of the WWTP.

Secondary control objectives should also be formulated (and met) for all conditions the lift station will experience during its entire projected life. Minimum flow rate conditions should be examined to provide acceptable pump control, prevent odor buildup, and supply the receiving unit process with adequate flow. Future flow rates should be reviewed to facilitate pump installation and ensure adequate wet well volume capacity. Control strategies should be developed for situations when one or more pumps are removed from service.

Another secondary control objective is to ensure the wet well volume provides adequate time between pump starts at critical flow rates. For a lift station with one active pump, the minimum cycle time (maximum cycle rate) occurs when the influent flow rate is 50% of the pump capacity. For lift stations with multiple pumps, all possible pump sequences should be evaluated to ensure sufficient time between pump starts. Sufficient time between starts allows a pump motor to cool after experiencing high starting inrush currents. For inadequately sized wet wells, automatically alternating pumps sometimes is used to provide enough cool-down time for a particular pump motor. "Soft-start" solid-state starters are becoming more commonplace and are required by some utilities to shave peak electrical demands. Because they produce less motor inrush current, they may have the added benefit of requiring smaller wet well volume.

INSTRUMENTATION

Instrumentation required to meet control objectives is listed in Tables 5.1 and 5.2.

Table 5.1 Minimum recommended lift station instrumentation

Device/process variable	Comments
Wet well level	Point or continuous measurement; for variable-speed, continuous measurement is a necessity; for multiple fixed-speed pumps, continuous measurement is common
High wet well level alarm	Ball float often used
Hand-off-auto hand switches	Operator interface
Thermal overloads	Motor protection

Table 5.2   Optional lift station instrumentation

Device/process variable	Comments
Check valve position	Monitor pump fail
Discharge isolation valve	To prevent surges, sometimes pump started/stopped against proven valve position
Low wet well alarm	Also often interlocked for redundant stop of pump; element subject to fouling
Phase monitors	Motor protection
Pump speed	Used with variable-speed drives
Flow rate	Used on important process streams; sometimes pump run times (with calibration factors) used instead
Ambient air monitoring (lower explosive limit, hydrogen sulfide)	Check requirements of National Fire Protection Association (NFPA) 820
Ventilation monitoring	Check NFPA 820 requirements
Bearing vibration	Pump protection
Overflow, dry well floor	Conductance probes often used
Thermistors, resistance temperature detectors	Motor protection
Pressure switches	Piping blockage
Moisture monitoring submersible pump seals	Diagnose defective seals
Monitor/measure shaft rotation	Alarm, feedback control
Elapsed run time	Assists in periodic maintenance and manual pump alternation
Alternators	See text
Indicating lights	On and off status, alarm conditions
Annunciator	Alarm conditions

Fixed- VERSUS VARIABLE-SPEED DRIVES

For a given lift station, pump drives can either be fixed- or variable-speed or a combination of the two. The selection of fixed- or variable-speed drives is an important decision, affecting the station's capital cost, its economy of operation, and the performance of downstream unit processes.

Relative to fixed-speed pumping systems, variable-speed systems typically have higher capital cost, require more maintenance, and are less "wire-to-water" efficient. Even though they are less "wire-to-water" efficient, variable-speed pumps are typically used where they might save energy in reducing dynamic head, that is, less piping friction loss at lower flow rates. Because they dampen the discharge flow rate, variable-speed pumps are also used where hydraulic surges adversely affect performance. Variable-speed pumps also reduce force main transients, which tend to extend piping, fitting, and fastener life.

Hydraulic surges in WWTPs are problematic, and variable-speed pumps are accordingly installed. Wastewater treatment plant process control problem areas related to hydraulic surges include disinfection, level control, and sludge blanket "burping" in secondary clarifiers.

Pumping applications with a relatively large percentage of dynamic head are good candidates for reduced energy costs with variable-speed pumping. However, it often costs less to use fixed-speed pumps in high static head and low dynamic head applications. Many lift stations fall into this category and are enough removed from WWTP processes to be better served with fixed-speed pumps.

Variable-speed pumps require less wet well volume because the wet well need not be sized to provide adequate time between pump starts. A smaller wet well is attractive in some applications because solid depositions and associated odors are fewer. Fixed-speed pumping stations in collection systems have been linked with odor problems.

Because wet wells of variable-speed pumps are typically smaller than those of fixed-speed pumps, savings in wet well construction costs for variable-speed systems are sometimes possible. Where excavation is deep or geotechnical site conditions difficult, savings using variable-speed pumps can be substantial. Though, in general, variable-speed pumping stations are more expensive than fixed-speed ones, this is not always the case.

Although only variable-speed drives have been discussed, variable flow pumping can be achieved by other methods, such as recycling discharge of fixed-speed pumps and installing throttling valves on the discharge of fixed-speed pumps. Because of their higher energy use, these methods are less common than variable-speed drives. However, they are sometimes used in retrofits and by users who prefer to avoid variable-speed drives altogether.

CONTROL SYSTEM HARDWARE

Choice of control system hardware depends on control complexity, availability of centralized plant control systems, user preference, and cost. With the advent of the microprocessor, choices of control system hardware abound.

A control system can either be dedicated to a particular lift station or can be part of the plantwide control system. Dedicated (or "stand-alone") systems typically physically reside at the lift station. They include conventional control relays, single-loop digital controllers, and programmable logic controllers (PLCs). Examples of plantwide control systems interfacing with lift stations include distributed control systems (DCS) and networked PLCs. In DCS and networked PLC systems, the logic executing device, or central processing unit (CPU), may or may not physically reside at the lift station. If the CPU does not reside at the lift station, field interface input/output (I/O) modules typically do.

CONTROL STRATEGIES

Control strategies (or algorithms) are a significant factor in successfully implementing control objectives of the lift station. Many different control strategies exist. A thorough analysis of lift station objectives is required to determine the most appropriate control strategy. Several common strategies are presented below.

TWO PUMPS, FIXED-SPEED DRIVES. Figure 5.1 shows typical start and stop levels for a lift station with two fixed-speed pumps. Assume incoming flow rate is less than lead pump capacity. Also assume there is no pump alternation. As the level rises above the "middle 2 (LSM2)" level, the lead pump starts and wet well level decreases. When the level falls below the "low (LSL)" level, the lead pump stops. As long as the incoming flow rate remains less than lead pump capacity, the lead pump continues cycling on and off. When the incoming flow rate increases beyond lead pump capacity, the wet well level continues to rise until the "high (LSH)" level is exceeded. Both lead and lag pump now run, and the wet well level decreases (assuming the incoming flow rate is less than the combined capacity of the two pumps). The lag pump stops when the wet well level falls below the "middle 1 (LSM1)" level. Note that if the lead and lag pumps are centrifugal, their combined capacity is less than additive because of increased dynamic head as the flow rate increases in shared piping.

Lead and lag pumps typically have separate stop-level setpoints, as shown in Figure 5.1. Use of a single stop-level setpoint is discouraged

---

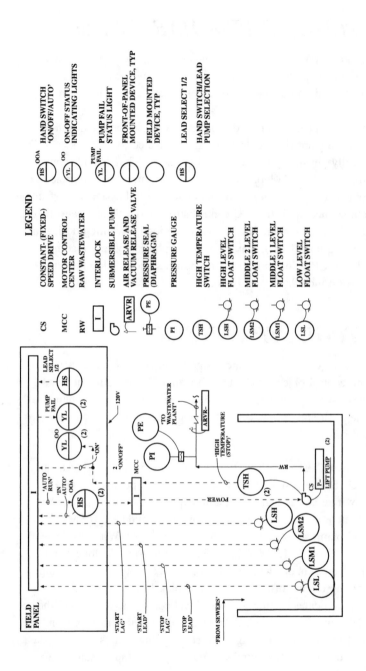

Figure 5.1 Example process and instrumentation diagram of a lift station with fixed-speed pumps (high high and low low alarms not shown) (WEF, 1993)

because this practice promotes larger discharge flow surges. Pump start levels are often set to ensure that the wet well level is less than the incoming line invert, preventing backup of wastewater in the incoming line. If wastewater is allowed to back up, there are potential benefits and risks. The main benefit is that the collection system typically has a larger capacity (volume) than the wet well. With an appropriate control strategy, the increased volume can smooth the lift station's discharge flow rate. Potential risks include possible upstream flooding, solids deposition in the incoming line resulting from reduced line velocity, and potential odor problems caused by solids deposition and increased retention time. Each case should be carefully studied before deciding to implement the strategy and, if implemented, to select the appropriate level setpoints.

CONTROL STRATEGY VARIATIONS, FIXED-SPEED PUMPS. As a result of readily available powerful microprocessor controllers, for example PLCs, practitioners continue to develop enhanced control strategies even for fixed-speed pumps. Such strategies offer reduced power use, reduced pump cycling, and smoother discharge flow patterns. In one strategy, lead and lag pumps operate using a total of two setpoints: start and stop. The two setpoints are set to make full use of the wet well volume. When the level falls below stop, the pump running the longest is stopped. The remaining pump is allowed to run for a timed period. At the end of the timed period, if the level continues to fall, it too is stopped. But if the level rises, the pump remains running. Pumps are started similarly.

In another strategy, the lift station consists of two small and two large pumps. The small pumps operate during low-flow periods, while the large pumps operate at high-flow periods. Matching pump capacity closely with incoming flow produces the benefits described above.

VARIABLE-FLOW PUMPS. Several techniques are typically used to achieve variable-flow pumping. These and other techniques are discussed in references by Skrenter (1988) and Sanks (1989). These techniques can control various final elements: discharge throttling valves; recirculation lines back to wet wells; and variable-frequency drives, magnetic couplings, and wound rotor motors. But as previously mentioned, discharge throttling valves and recirculation lines are energy inefficient. Wound rotor motors are also inefficient and are becoming obsolete.

Load sharing (Sanks, 1989) is the most commonly used variable flow control strategy. It is applicable for variable-speed pumping and discharge valve throttling. In load sharing using lead and lag variable-speed pumps, both run at the same speed. In the non-load-sharing strategy, during certain times the lead pump may run at constant maximum speed, while the lag pump speed is variable. Load sharing is a more efficient strategy than non-load-sharing. However, a non-load-sharing strategy allows only one variable-speed drive to control two pumps. Once at maximum speed, the lead pump is transferred to the drive, which is then switched to adjust the lag pump speed.

It is important to decide the appropriate strategy to adjust drive speed. Broadly speaking, drive speed can controlled either to maintain a level setpoint or to vary in proportion to surface level within a level band.

Level-setpoint control has been successfully implemented in many systems but caution is advised. Maintaining tight level-setpoint control does little to dampen pumped flow variations; it may also cause wild oscillations in drive speed as the drive strives to maintain the level setpoint. From a mass balance view, all wet well storage is eliminated if the level remains perfectly constant. This allows no smoothing of the incoming flow.

In a level band approach, a band is assigned to each sequence step, for example, lead and lag pumps running together. Drive speed increases as the level rises in the band. If the incoming flow rate exceeds the step's pumping capacity, the level eventually rises past the step's band. This triggers a new step, for example, lead, lag 1, and lag 2 pumps running together with a new level band. To conserve vertical wet well height, bands of different steps often overlap. The level band approach tends to smooth the discharge flow rate. Band size is application specific, but a 0.6-m- (2-ft-) high band (for each step) may be considered typical.

VARIABLE-SPEED DRIVES. For incoming flow rates below lead pump capacity, the lead pump speed is modulated to maintain either a level setpoint or a level band. The incoming flow rate may become so low that it is less than the discharge flow rate of the lead pump at minimum speed. During such times, the lead pump is cycled on and off at minimum speed. (Variable-speed pumps must be run above a minimum speed to prevent head buildup in the pump motor and, in some cases, the pump itself.)

When the incoming flow rate exceeds lead pump capacity, the lag pump is started at a speed equivalent to approximately 50% capacity. Lead pump speed is decreased to match lag pump speed. Logic is provided to prevent unnecessary "bumps" in station discharge flow rate. In load sharing, the speeds of both pumps are modulated to be equal. With both pumps at the same speed, the strategy strives to maintain either a level setpoint or an acceptable level band. When the incoming flow rate is again within lead pump capacity, the lag pump is stopped.

FIXED- AND VARIABLE-SPEED PUMPS. Many large lift stations use a combination of fixed- and variable-speed pumps to obtain the process benefits of variable flow rate and the economic benefits of fixed-speed pumps. Control strategies can become complicated. Fixed- and variable-speed pumps exhibit discharge flow rate discontinuties, or gaps, if not properly designed. Such gaps may adversely affect performance of the downstream unit process. At certain incoming flow rates, gaps may also create excessive and harmful pump cycling.

Often, designers resort to a larger variable-speed lead pump and a smaller fixed-speed lag pump. This increases lift station complexity. Such an arrangement reduces gaps as it provides an overlap in pumping capacities of successive sequence steps. The designer should provide a discharge flow rate at the start of the new step that is less than the discharge flow rate at the end of the preceding step. For instance, the combined discharge flow rate of the fixed-speed lag pump plus the lead pump at minimum speed (the new step) should be less than the discharge flow rate of the lead pump at maximum speed (the previous step).

STANDBY PUMPS. Except for minor applications, lift stations are provided with at least one standby pump for increased reliability. Overloading the downstream unit process with an excessive process flow rate should be evaluated. If overloading is unacceptable, the control strategy should limit the number of pumps in operation. If necessary, similar logic should be provided to prevent overloading the power distribution system.

VARIABLE-SPEED PUMPS, EQUALIZATION BASIN. In most wet wells, pump control is linked to level. However, in equalization basins with large wet wells, the designer has the option of using discharge flow rate as the primary control variable (level serves as a secondary control variable, disabling pumps on a falling low level). Guaranteed flow rates and smooth flow patterns can be delivered to the receiving unit process. Pump speed is controlled to maintain a discharge flow rate setpoint, which is operator-entered. In simple strategies, the number of pumps to run is decided by the operator. In more complex strategies, pumps are automatically started and stopped. A common approach is to start the next pump in the sequence when the feedback controller output remains high for a sustained time.

*P*UMP ALTERNATION

Control strategies also alternate pumps. The objective of alternation is to evenly distribute pump wear. In the opinion of some, however, equal pump wear promotes simultaneous pump failure, and alternation should be avoided.

Various alternation hardware are available. These include electromechanical alternators, electromechanical stepping switches, microprocessor-based alternators, solid-state alternating relays, lead and lag selector switches (for manual alternation), and manual alternation by plug and jack systems. Programmable logic controllers and DCS controllers can also be programmed to provide alternation.

Alternation strategies range from simple to complex. A common simple strategy is to alternate lead and lag pumps on falling low level when both pumps stop. Complex alternation strategies include first-on/first-off and last-on/first-off. The first-on/first-off strategy stops the pump with the longest cycle run time, and starts the pump with the longest rest time. Another complex strategy provides uneven run time on pumps.

For lift stations routinely visited, an appropriate strategy may be to keep pumps in a fixed sequence, which is periodically changed by an operator turning a selector switch.

Common ways to automatically initiate alternation include reaching a low level when all pumps are off, at periodic real-time intervals, or when the run time of a pump for its present cycle exceeds a preset value.

ABNORMAL OPERATING CONDITIONS

Abnormal operating conditions must be accounted for in control system design. The control strategy should provide an orderly start-up after a power outage. To reduce hydraulic transients and power distribution overloads, pump start times are often staggered. Protective control circuits should also be analyzed to verify that pumps automatically restart after an outage.

For a control system using a PLC, its CPU and I/O peripherals should be able to resume operation after an outage or "brownout." For certain PLCs, some reconfiguration is required.

Programmable logic controllers benefit from conditioned power. Some but not all designers choose an uninterruptible power supply (UPS) to provide conditioned power. If a UPS is used, the control strategy must be reviewed to ensure proper start-up after an outage. For example, a UPS-backed PLC may sense a false pump failure during the outage. Once power is resumed, the false failure may prevent the pump from restarting.

Often, for its own protection, a pump is shut down at an abnormally low level. To restart the pump, manual reset may be problematic if an operator is not present to invoke the reset. This could cause the lift station to overflow when incoming flow resumes. Automatic restart as the level rises higher may be more appropriate. The resulting level deadband allows the pump to remain idle and decreases cycling to an acceptable amount. Excessive cycling can damage the pump and its associated piping system.

REFERENCES

Sanks, R.L., Ed. (1989) *Pumping Station Design.* Butterworth-Heinemann, Stoneham, Mass., 353.

Skrentner, R.G. (1988) *Instrumentation Handbook for Water and Wastewater Treatment Plants.* Lewis Publishers, Chelsea, Mich.

Water Environment Federation (1993) *Design of Wastewater and Stormwater Pumping Stations.* Manual of Practice No. FD-4, Alexandria, VA.

Chapter 6
Flow Splitting

It is often necessary to split flows among multiple units (such as several primary settlers) or different unit operations (such as "new" and "old" wastewater treatment plants [WWTPs]). In many cases, it is difficult to split flows accurately over commonly encountered ranges of flow (often greater than 10:1). Many types of flow splitting mechanisms have been used, including static (nonmodulating) hydraulic structures, manually adjusted weirs and gates, and automatically controlled weirs and gates with a flow measurement for feedback control. The use of automatically controlled mechanisms allows accurate flow splitting with minimum head loss.

*S*TATIC STRUCTURES

Static structures are designed using simple weir, Parshall flume, or orifice formulas. The equations themselves are based on mass and momentum equations with ideal inlet and outlet conditions. Because most flow splitting structures do not satisfy these ideal conditions, the mass and momentum equations may lead to some inaccuracies and not perform according to design. Inaccuracies during construction can also prevent static structures from operating properly. Therefore, the designer must make an early decision regarding the degree of accuracy required of the flow splitting device. Static structures generally require more head loss than other flow splitting mechanisms.

MANUALLY ADJUSTED WEIRS, GATES, AND VALVES

Flow splitters with manually adjusted weirs, gates, and valves may also have difficulty operating well. These structures depend on equalizing the hydraulic head loss between units to achieve equal (or design) flow splitting. Either because of design configuration or construction inaccuracy, there are always differences in hydraulic resistance among the different paths. These differences in resistance are nonlinear functions of velocity. For instance, the head loss down a pipe is a function of the velocity to the 1.5 power (Manning equation). Therefore, the flows can be equalized only at a specific flow rate. As the flow moves away from this operating point, the flow split becomes unequal (poorer control) unless the weir or gate is manually adjusted. Manually controlled flow splitters may require frequent operator attention during times of greatly changing flows.

AUTOMATICALLY CONTROLLED FLOW SPLITTERS

Several strategies have been devised to control flow splitting dynamically with modulating control elements. Two successful strategies and one unsuccessful strategy are described below. The unsuccessful strategy is described because it has been used (unsuccessfully) on numerous projects in the past. Minimum and optional instrumentation needs for flow splitting by the two successful methods are given in Tables 6.1 and 6.2.

Table 6.1   Minimum recommended flow splitting instrumentation

Measurement	Comments
Flow rate	One needed for each effluent channel or pipe
Control element (valve or gate)	One needed for each effluent channel or pipe
Influent level	Needed only for influent channel level splitter

Table 6.2   Optional flow splitting instrumentation

Measurement	Comments
Position indicator	One needed for each final control element
Influent flow rate	Used for flow meter verification

SUCCESSFUL FLOW SPLITTING STRATEGY NO. 1. Figure 6.1 shows a piping network with flow splitting among three units. Equivalent open channel and pump speed control examples can also use the identical strategy. For the purpose of this example, it is assumed that equal flow splitting is desired. However, it is rather easy to implement a ratio control to divide the flow unequally to account for unequally sized units.

The general strategy is to turn off control to one unit and set the control element to a fixed position (for example, 90% open). The measured flow from that unit is used as the controller setpoint for the other control loops. This simple strategy works well until an external constraint is violated (for example, influent level or pressure is too high) or one of the controllers cannot meet its setpoint. In these cases, the fixed control element position is adjusted until good control is achieved again. For instance, if one controller has its control element all the way open and still is below the flow setpoint, the fixed control element should be closed some. Likewise, if the influent pressure is too high, the fixed control valve should be opened some. The amount that it should be opened or closed may vary from less than 0.5% to more than 5.0% of the range and depends on the specific application.

SUCCESSFUL FLOW SPLITTING STRATEGY NO. 2. Another successful flow splitting algorithm that works especially well when the influent level is a constraint is shown in Figure 6.2. The influent level (or wet well or channel level) is used to generate a flow setpoint for each of the units. A number of algorithms can be used for this purpose ranging from a feedback controller that maintains a near-constant level to a simple linear function or second- or third-order function in which level may vary significantly. Usually, a simple linear function (straight line) is used to generate a flow setpoint of zero (or near zero) at the lowest anticipated level and the maximum flow capacity of the control element at the highest permissible level. It is also possible to add ratio controllers ahead of the flow elements to achieve unequal flow splitting for differences in unit capacities. This algorithm is not adversely affected by minor inaccuracies between the flow meters.

UNSUCCESSFUL FLOW SPLITTING STRATEGY. A common unsuccessful control strategy uses the same instruments and control elements described above as well as a total influent flow rate meter not shown in Figure 6.1. The setpoint for each individual flow loop is derived by dividing the total flow rate by the number of units. This strategy fails because each flow meter is not "perfectly" accurate and each flow loop cannot "perfectly" control the flow. The control valves will tend to close completely if the sum of the individual flow meters is greater than the single influent meter. Likewise, they will tend to open all the way if the individual flow meters sum to less than the single influent meter. This control strategy cannot be successful and is not recommended.

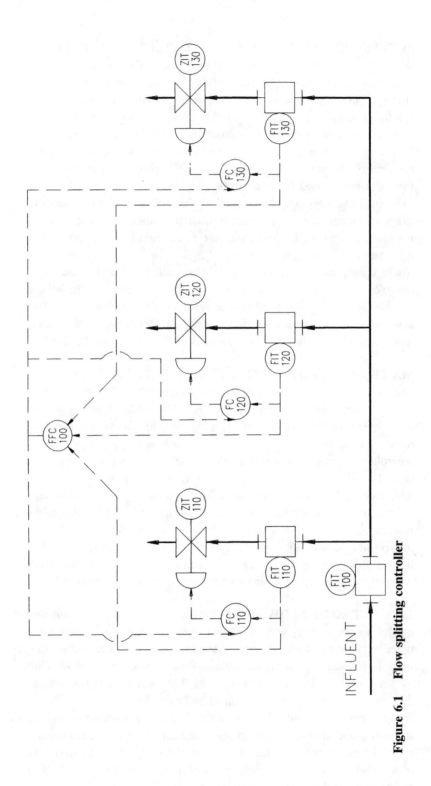

Figure 6.1 Flow splitting controller

Automated Process Control Strategies

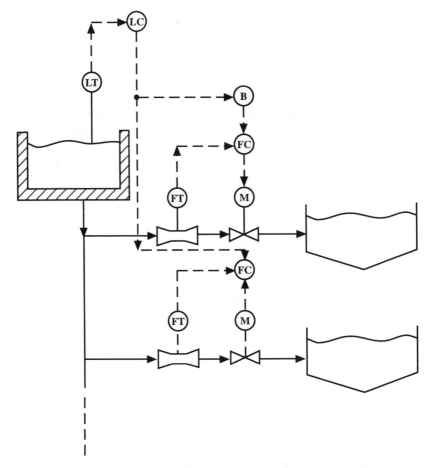

Figure 6.2   Level to flow cascade control loop for flow splitting

*F*LOW METER VERIFICATION

Because the accuracy of flow splitting is dependent on an accurate flow measurement, the designer should pay particular attention to all meter installation details. Excellent design checklists for many types of flow meters are readily available (Skrentner, 1989). Even for perfectly installed flow meters, however, there is a need for periodic calibration or flow rate verification. Several types of flow verification include the use of redundant flow meters or experimental tests using the drawdown method or various dye dilution methods. The flow meter installation should conveniently provide for one of these types of flow meter verification.

FINAL CONTROL ELEMENT ACTUATORS

The time constant for flow rate through a WWTP is typically in the range of 5 to 10 minutes. Depending on the variability of the influent flow rate, therefore, it may be necessary to make adjustments to the flow splitter's final control elements at time increments ranging from less than 1 minute to several hours. Control periods of from 1 to 5 minutes are common. The final control element actuators, consequently, should be capable of near-continuous operation. Pneumatic and hydraulic actuators are generally preferred over alternating current motor actuators.

REFERENCE

Skrentner, R.G. (1989) *Instrumentation Handbook for Water & Wastewater Treatment Plants*. Lewis Publishers, Boca Raton, Fla., 33.

Chapter 7
Preliminary Treatment

CONTROLS

Most equipment manufacturers can supply controls for operating the preliminary treatment system. These controls vary widely in capabilities, ranging from simple on–off switches to complex control panels with motor load sensors, ammeters, motor-reversing actuators, and alarm signals. Declining costs may make remote video monitoring an affordable option.

The types of controls used at a given facility should reflect the degree of automation of the rest of the system. At small wastewater treatment plants (WWTPs), which typically do not have a central control system, a local hand-off-auto switch should be in the vicinity of the unit. At larger WWTPs with a central control system, more complex controls may be necessary to monitor the system for proper operation and alarm conditions. Supervisory control and data acquisition (SCADA) systems will also monitor items such as running time and will alert the operators of scheduled maintenance requirements.

SCREENING AND CONVEYANCE

Wastewater debris is typically screened from the influent waste stream before primary and secondary treatment to prevent clogging and damage to downstream equipment. Screens are constructed of a series of bars with openings typically varying between 10 and 30 mm. As wastewater flows through the screen, debris larger than the opening size is strained out and accumulates on the bars. This debris accumulation causes head loss across the screen and tends to raise the upstream water level. At some point before the screen channel overflows, the debris must be removed by a rake mechanism. Debris removal is typically done periodically, rather than continuously, to minimize mechanical wear and the consequent maintenance. The rake mechanism then dumps the accumulated debris (called screenings) on a conveyor for conveyance to storage, treatment, and/or disposal.

The screening equipment must be interlocked with conveyors and other downstream equipment such as a screening press. Conveyance equipment requires local start/stop ability and an emergency stop for personnel safety. Remote start/stop ability is often provided along with motion sensors for status.

SCREENING CONTROL STRATEGIES. Two of the main control objectives for screening are to initiate the start of the cleaning cycle and to protect the screening equipment from mechanical damage that could be caused by jamming of the mechanism. Protection can be provided by motion sensors or, more commonly, by motor torque overload relays. These relays first alarm a high-torque condition, then shut down the equipment if a higher torque is sensed. Such basic instrumentation is typically provided by the manufacturer of the screen, along with other status instruments.

There are several strategies commonly used for initiating the screen-cleaning cycle. These strategies include elapsed time, high influent channel level, cumulative flow, and differential level across the screen.

Elapsed Time. The use of elapsed time to initiate a screen-cleaning cycle is perhaps the simplest of all strategies. Timers are set to initiate the cleaning cycle on a periodic basis (for example, 30 minutes). This strategy is simple and inexpensive to implement and requires only one or two timing relays. Because no measurements of screen performance are taken, the cycle time must be set conservatively for worst-case conditions. Therefore, on average, the screen-cleaning mechanism cycles more often than necessary. Mechanical maintenance will be greater than with other strategies, and performance cannot be guaranteed.

High Influent Channel Level. Screen cleaning may be initiated by a single influent channel high-level measurement. The instrument for this measurement is often an inexpensive, discrete point measurement device. As debris accumulates on the screen and head loss increases, the water level increases in the influent channel. When this level passes the setpoint, the cleaning cycle is initiated. Water level will also increase with higher influent flow rates. The simple measurement device, however, cannot differentiate between level changes resulting from flow rate and those resulting from debris accumulation. Therefore, the level setpoint must be set conservatively. While somewhat better than the elapsed-time strategy, on average, the screen-cleaning mechanism cycles more often than necessary, resulting in excessive mechanical wear.

Cumulative Flow. Another strategy for cleaning screenings is to measure the total flow going through the screen and initiate a cleaning cycle after a given quantity of wastewater has passed through. This strategy has some appeal because it results in the desired characteristic of cleaning the screen more often when the flow rate increases. However, the strategy makes the inherent assumption that the accumulation of debris is constant with time. In reality, the amount of debris at most WWTPs varies greatly. It is typically the highest during times of high flow, when accumulated debris in the collection system is swept to the WWTP.

Differential Level. A highly desirable strategy for controlling screen cleaning is the use of differential level across the screen. While differential level is both a function of blockage and flow rate, it is an excellent determinant of when the screen must be cleaned. Differential level is commonly measured by a differential pressure cell with two bubbler systems or by a dual-headed sonic level instrument. The use of differential level minimizes the number of cleaning cycles and subsequent maintenance costs while guaranteeing screen performance. A process and instrumentation diagram of this strategy was presented as Figure 4.6.

UNEXPECTED CONSEQUENCES OF SCREEN-CLEANING CYCLES.

If two or more parallel screens have a common influent channel but separate discharge channels (such as to ''new'' and ''old'' WWTPs), they can function inadvertently as a poorly operating flow splitter (see Figure 7.1). It is possible that the two screens will clog at unequal rates. Whenever one of the cleaning cycles is initiated, the newly clean screen receives most of the influent flow. As that screen clogs and the other screen is cleaned, the flow will shift towards the other screen and the second WWTP. This situation results in flow rates to each of the two WWTPs going from almost zero flow to the total plant flow on a rapid basis. Overall WWTP performance would be greatly compromised.

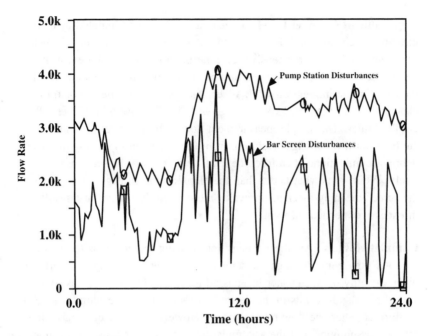

Figure 7.1   **Flow variations to wastewater treatment plant before elimination of bar screen hydraulic disturbances ($\circ$ = total flow, m^3/h, and $\square$ = plant no. 2, m^3/h; plant 1 flow = total flow − plant 2 flow)**

Relatively simple controller revisions can eliminate the alternating flow rate change that would result from this situation. However, this illustrates the fact that even simple control strategies can have a major impact on the performance of a WWTP and should not be neglected in design or operation.

GRIT REMOVAL

Any type of grit removal system will have a monitoring/control system. This may range from the manual monitoring/control system, which is based on operator observation and measurement, to a system of instruments that provides the monitoring/control system for the grit removal process. This section describes those monitoring/control systems that may be present in the system and their operation and maintenance for best performance.

LIQUID LEVEL. A liquid level system may be as simple as numbers on a gauge that the operator reads visually, or as complicated as an electrical or pneumatic level sensor. These devices may have a local indicator or may send a signal to a remote instrument display or to a computer data-logging or control system.

Liquid level sensors assist the operator in knowing when to bring grit tanks on or off line. These sensors should be located in stilling wells, and bubbler sensors should be provided with the ability to physically remove any debris on or in the tubes.

DENSITY METERS. Larger WWTPs that are fully instrumented may include meters that measure the density or suspended solids in collection areas or flowing in a grit line. Such devices can assist an operator in transferring grit so that lines do not become plugged.

FLOW SWITCHES. Flow switches detect the movement, or lack of movement, of fluid in pipes. Such devices should be located in stilling wells when used and can provide valuable information to an operator hydraulically transferring grit in pipes.

ROTATION STOPPAGE SWITCHES. Grit removal systems that use chain-and-flight scrapers, grit-dewatering screws, pumps, bucket elevators, or similar rotating equipment may be equipped with these devices to sense a jam or shear pin failure.

HIGH-TORQUE SWITCHES. An alternative to rotation stoppage switches, these devices may be triggered by motor current draw and, therefore, warn of impending problems so that an operator can make adjustments in a timely manner.

Chapter 8
Primary
Sedimentation

Primary sedimentation refers to the gravitational separation of particulate organic and inorganic solids from influent wastewater. Suspended material must be sufficiently greater in size (at least 50 μm) and density than water that gravitational separation is possible within the limitations of the tank's hydraulic retention time. The process typically removes 40 to 50% of the influent suspended solids. These solids are responsible for 20 to 40% of the influent biochemical oxygen demand (BOD). Reducing the BOD load to the secondary process also reduces the amount of required air. A reduced air requirement results in less energy consumption and overall cost. An additional function of the primary tanks is to remove floating grease or scum. Scum is collected and pumped periodically depending on the scum level within the hopper or sump. The clarified liquid typically receives secondary treatment. The thickened solids also typically receive further treatment and thickening or dewatering.

The design of primary sedimentation tanks is governed by the surface overflow rate and the hydraulic retention time. Surface overflow rate determines the settling velocity of the smallest particle that will be completely removed according to discrete (type 1) sedimentation. Hydraulic retention time is important because flocculant (type 2) sedimentation also generally occurs in primary settlers. Based on these two parameters, the geometric characteristics of the settler can be formulated. Because of the intermittent nature of the settled solids removal, extra capacity is often added for storage in the sedimentation tank.

Control Objectives

Because the size of the settling tank is fixed in design, the primary variables that can be manipulated are the number of primary settlers on line and the withdrawal rate. The number of primary settlers on line is typically adjusted only on a seasonal basis, although some wastewater treatment plants (WWTPs) change more frequently.

The withdrawal rate affects the amount of time the solids spend in the settler, the sludge blanket level, and the solids concentration. Generally speaking, a low withdrawal flow rate results in an increased solids retention time (not to be confused with sludge age), higher sludge blanket levels, and an increased underflow concentration. Likewise, a high withdrawal flow rate results in decreased solids retention time, a lower sludge blanket level, and a more dilute solids concentration.

The primary settler itself can typically operate successfully over a wide range of withdrawal flow rates. However, too long of a solids retention time can result in several undesirable phenomena. If retained too long, primary solids may undergo anaerobic decomposition, which produces gas bubbles. These bubbles tend to adhere to clumps of solids and cause the solids to rise to the settler surface. These floating clumps, sometimes called "rafts," degrade the performance of the settler because they end up in the primary effluent. Anaerobic decomposition may also cause some particulates to form soluble BOD, again negating some of the benefits of primary sedimentation. Additionally, if the solids become too thick, they may be difficult to pump.

Conversely, if the withdrawal flow rate is too high, the solids concentration will be too dilute. Such dilute solids may adversely affect the operation or efficiency of downstream processes. Note that some facilities may follow primary sedimentation with a subsequent thickening process that reduces the consequences of dilute primary sludges. In most cases, however, a balance must be reached between thickening enough for downstream processes and the negative results of overthickening.

Control Strategies

For most WWTPs, the sludge withdrawal flow rate is controlled by running fixed-speed pumps for a fraction of the time. For instance, a WWTP may pump for 5 minutes of every hour. A few WWTPs have variable-flow pumps.

MINIMAL CONTROL STRATEGY. The minimal control strategy requires the manual start and stop of withdrawal pumps and collection equipment such as rakes, screws, and valves. The operator makes manual measurements and manually operates the equipment to maintain the required sludge blanket level and underflow concentration in each tank. Scum pumping is based on the level of the scum in the collection hoppers or sumps. The efficiency of manual control depends greatly on the skill and diligence of the operator.

MID-LEVEL CONTROL STRATEGY. Perhaps the most commonly used control strategy consists of simple timing of collected and pumped solids and scum. The timers are manually set or programmed into the controller for each tank and pump. The length of time required for pumping is based on frequent sample data of the solids and scum depth in the tank and/or the solids concentration. The addition of flow rates to the timing pump strategy will assist the operations staff in determining the total amounts being withdrawn from the tanks and in solids inventory. Depending on the placement of the flow meter(s), the pumping sequence for each tank could be monitored for withdrawal rates. Single-speed collection equipment and pumping is often used for withdrawal, although it may lead to some operational difficulties. Often, thinning of the sludges will be apparent, and greater tank capacity will be needed for subsequent processes such as blending tanks and digesters. Variable-speed collection equipment and pumping have the added benefit of adjusting to WWTP influent flows, varying tank conditions, and piping limitations downstream. The use of programmable timers or controllers, in conjunction with manual sampling of the solids and scum depth and/or solids concentration, would be the minimum requirement for an automatic primary sedimentation control strategy.

Use of Programmable Logic Controllers. Programmable logic controllers, or PLCs, coupled with variable-speed collection equipment and pumping, can offer greater control in monitoring the settling and withdrawal processes. Different operation scenarios can be preprogrammed into the PLC to account for storm flows, tanks out of service, and emergency bypass situations. This type of programming allows faster response times to changing conditions as well as less disruption to the WWTP processes. Operator intervention of the automatic control strategy in the case of emergency situations or maintenance is easily accomplished using a PLC and operator terminal.

This mid-level solution to automating primary sedimentation control will still require some manual input from the WWTP operations staff concerning the depth and/or concentration of solids. However, these setpoints can quickly be entered into the process control strategy by the operator to achieve appropriate adjustments.

ADVANCED CONTROL STRATEGY. The minimal and mid-level types of control strategies require significant manpower to collect data. Samples of sludge level and solids concentration may have to be taken as often as every 2 hours or as infrequently as once per day, depending on WWTP process conditions. There may be considerable variance in the quality of the data, depending on the skill and diligence of personnel involved with data collection and analysis. The data collected must be accurate for the results of any control strategy to be effective. For small WWTPs that have few primary tanks, manual data collection may be acceptable. For larger WWTPs with multiple batteries of primary tanks, the manual data collection and correlation can become burdensome.

An advanced computer or PLC control strategy could use a single-point or multipoint sludge blanket depth measurement to determine whether each tank has sufficient solids for pumping, a solids concentration analyzer, and a solids flow meter. The sludge blanket level is used to control the pumps. The pumps start as a function of time and stop after a time delay or when the sludge blanket level in the tank reaches a low value. A timing program sequentially starts the sludge pumps at preset time intervals to reduce downstream piping pressures.

The pumps should be sequenced such that only a given number of pumps (one or two) per battery of tanks is started at a time. Scum pumps are typically single speed. The variable-speed sludge pumps start at a predetermined low speed to reduce wear on the pumps. After a preset time delay, the sludge pumps ramp up to the determined setpoint speed. The speed would be controlled by the level of sludge and/or concentration being pumped. The number of pumps running and the control setpoint for optimizing the feed rate to the next process downstream could also be set by the sludge flow meter(s). The pumps would stop pumping if the solids monitor detected a concentration that was too dilute or the sludge blanket level instrument indicated that a low level of sludge existed within each tank.

A corollary to this strategy could be the addition of minimum pumping time once the pump is started to allow for settled solids contained in the pump discharge piping to be cleared before accurate solids density data are reported. The withdrawal for each primary tank would be individually controlled and coordinated with all the tanks in service. This strategy demands that the many measured variables be accurate or at least repeatable and that the measuring equipment be in service.

It is commonly acknowledged that the present state of the art of instrumentation for high solids concentrations is far from perfect. These instruments are often expensive to purchase, install, and maintain. Fouling of the sensors, plugging, and grease coating of the monitoring points may occur frequently. However, these problems can be minimized if proper design is implemented for maintenance of this type of equipment. The Instrumentation Testing Association has published full-scale WWTP results for high-range solids instruments (ITA, 1993). Typical instrumentation requirements for primary sedimentation are listed in Table 8.1.

Table 8.1   **Instrumentation requirements for primary sedimentation (WEF, 1993)**

Measurement	Comments
Minimum recommended primary sedimentation instrumentation	
Sludge flow rate	Totalize; provide adequate cleaning
Hand-off-auto switch	Operator convenience; manual operation
Timing relays	Time-based operation of sludge pumps; may be software timers in a programmable logic controller
Optional primary sedimentation instrumentation	
Sludge solids concentration	Subject to fouling and interferences; may require extensive maintenance; may be performed manually
Sludge blanket level	Subject to fouling and interferences; may require extensive maintenance; may be performed manually

The benefits and limitations of automated monitoring must be weighed carefully. Most large WWTPs, however, will use some form of automated monitoring and advanced computer control strategy for primary sedimentation, scum collection, and pumping because the benefits far outweigh the costs.

REFERENCES

Instrumentation Testing Association (1993) *Performance of High Range Suspended Solids Analyzers For Wastewater Treatment Applications.* Washington, D.C.
Water Environment Federation (1993) *Instrumentation in Wastewater Treatment Facilities.* Manual of Practice No. 21, Alexandria, Va.

Chapter 9
Activated-Sludge Treatment

*P*ROCESS DESCRIPTION

Activated-sludge treatment is a biological process commonly used for treatment of municipal and industrial wastewater. The process consists of a biological reactor (or reactors) followed by secondary sedimentation basins, as shown in Figure 9.1. In the biological reactor, the influent wastewater is

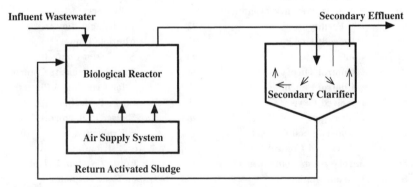

Figure 9.1   Process diagram for activated-sludge treatment system

brought into contact with a mixed culture of microorganisms—the mixed liquor—suspended in concentrations generally ranging from 500 to 5 000 mg/L. In the biological reactor, the organic material from the influent wastewater is either consumed by the microorganisms or entrapped in the floc formed by the mixed culture of microorganisms (biomass). The air supply system provides both the oxygen necessary for biological reactions and the energy required to keep the solids in suspension. The floc, including the biomass and the entrapped nutrients, is separated from the liquid in the secondary clarifier. The secondary effluent overflows from the top of the clarifier, and the concentrated solids are returned to the biological reactor as recycle or they are wasted. The key part of the process is recycle (return activated sludge, or RAS), which allows for the buildup of the mixed liquor concentration high enough to greatly speed up the degradation process that would occur in nature given enough time.

A number of specific process configurations and schemes were developed from the basic activated-sludge process. These configurations may vary in terms of the air supply mechanism (air versus pure oxygen), or with respect to the contact pattern between the influent wastewater and the biomass (step feed, contact stabilization, plug flow). Some process modifications include anaerobic and/or anoxic phases combined with aerobic zones.

*I*NTERACTIONS

An understanding of interactions plays an important role in our understanding of the activated-sludge process. Although operating engineers have been aware of the importance of the interactions for some time, interactions have not been studied extensively in the past. To investigate interactions on a quantitative basis, it is necessary to have a comprehensive model that includes several process units. Developing such a comprehensive model is a complex task because the mathematical models for individual process units are still under development and model validation is still difficult and incomplete. To make good overall operational choices, the following types of interactions must be understood and considered.

INTERACTIONS BETWEEN PROCESS UNITS. Interactions between process units are important both in design and in operation. In an activated-sludge wastewater treatment plant (WWTP), the design and operation of the biological reactors have a great effect on the secondary clarifiers because they are physically linked through wastewater and RAS flows. For example, if the aeration basins are designed smaller, the concentration of the mixed liquor suspended solids (MLSS) would have to be higher to maintain the same food-to-microorganism (F:M) ratio or solids retention

---

time (SRT). The increase in the MLSS would result in increased solids loading to the secondary clarifier; therefore, the clarifier would have to be larger to maintain the same solids-loading-to-surface ratio.

The performance of the secondary clarifier is also strongly affected by the settling properties of the MLSS, and the conditions under which biological growth occurs have been recognized to have a pronounced effect on settling properties. In addition to achieving the adequate level of pollutant removal, the aeration basin must produce solids with good settling properties.

An example of the cost interaction between the process units is presented by Stenstrom and Andrews (1980). Stenstrom and Andrews modeled the activated-sludge process and explored the effect of different operating strategies on the overall cost of operation. The simulation results confirmed the effect of operational interactions and were demonstrated in the following example. When the SRT of the system was increased, the oxygen requirements increased, while the digested solids production and the methane production in the anaerobic digester decreased. For the particular WWTP studied, this meant that more money had to be spent to supply oxygen, while at the same time, less methane was available for sale to the local gas company. While for this particular case the overall costs increased, Stenstrom and Andrews showed that the situation might be different for a WWTP in which energy is less expensive and/or the cost of solids disposal is higher. In any case, an understanding of the interactions is required to find the optimal overall strategy.

INTERACTIONS BETWEEN DESIGN AND OPERATION. The importance of the interactions between design and operation has long been recognized by WWTP operating engineers. To accommodate a change in operating conditions, operators need the flexibility to manipulate the WWTP's processes and adapt to changing conditions in an optimal way. To allow for such manipulation and build in the necessary flexibility, it is important to consider fully the dynamics of WWTP operation during the design stage. One example of such interaction is dissolved oxygen (DO) control. While controlling the DO setpoint can result in energy savings, the control strategy can sometimes encounter constraints imposed by the design. If the tank volume is too big, the energy necessary to keep the solids suspended may exceed the energy required to supply enough oxygen to adequately support the biological reaction. In addition to the tank design, the design of the air supply system can affect the DO control system as well; for example, the DO control strategy can be limited if appropriate turndown capacity or flexibility of the air supply system is not incorporated into the design.

INTERACTIONS BETWEEN CONTROL LOOPS. Interactions between control loops can occur when the process is manipulated simultaneously by several subsystems, such as the MLSS controller manipulating

SRT in the system by controlling the volume of wasted activated sludge; the DO controller manipulating air (or oxygen) distribution to biological reactors; the influent pumping station controls regulating the flow into the process; and/or the blower control principally regulating the header pressure setpoint.

INTERACTIONS BETWEEN SYSTEMS. Interactions also exist between the activated-sludge process and other connected systems, such as solids processing or the collection systems network. For example, the available capacity in some collection systems networks might be used to equalize the flow to the WWTP, thus allowing a logical connection between the control systems for the sewers and the WWTP.

*A**ERATION CONTROL*

The major objectives of aeration control are to ensure that the supply of oxygen meets the dynamic variations in process biomass oxygen demand, to control air delivery and oxygen transfer effectively to minimize aeration energy costs, and to minimize manpower to accomplish these goals. Automated aeration control is the manipulation of the aeration rate by computer or controller to match the dynamic oxygen demand and maintain a desired residual or setpoint mixed liquor DO concentration.

BENEFITS OF AERATION CONTROL. The benefits of implementing an aeration control strategy in a WWTP are ensured integrity and uninhibited operation of the process, increased reliability in meeting plant discharge requirements, and reduced process costs. If properly applied, aeration control offers the potential for significant cost savings.

Process Implications. Maintenance of an inadequate residual mixed liquor DO concentration can inhibit biological activity and contribute to process problems such as sludge bulking, poor flocculation, and inhibition or loss of nitrification. Conversely, excessive aeration results in excess energy costs. Improved process reliability, nitrogen removal efficiency, solids settleability, and effluent quality have been attributed to automated DO control.

Economic Considerations. Aeration energy consumption typically represents 50 to 90% of the total energy demand for activated-sludge plants (Wesner *et al.*, 1977). The incorporation of effective automated control of the aeration process can result in considerable energy savings. However, as the complexity of a control system increases, labor costs can increase. The design goal is to select the optimal control system, that is, one that achieves satisfactory treatment at minimal total cost.

MANUAL CONTROL. Variations in oxygen demand make it difficult, if not impossible, for an operator to manually manipulate air flow rates and air distribution to maintain desired mixed liquor DO concentrations throughout a sustained operating period. This is true even for WWTPs with well-designed, flexible aeration systems. Therefore, manually adjusted aeration systems are typically operated at fixed air flow rates and distributions. Changes are initiated once or twice daily at best, and often on a weekly or seasonal basis only. Air flow is manually fixed at a rate high enough to satisfy the oxygen demand anticipated during peak loading periods. This practice results in unnecessary and costly excess aeration during extended periods of reduced loading.

AUTOMATED CONTROL. Automated DO control is the best way a well-designed aeration system can be manipulated effectively to satisfy biomass oxygen demand, minimize operational problems associated with inadequate or excessive aeration, and minimize aeration energy consumption. Generally, the potential aeration energy savings achievable by automatic aeration or DO control is 25 to 40% (Flanagan and Bracken, 1977, and Stephenson, 1985). Potential savings are WWTP-specific and depend on loading characteristics, configuration and process hardware design, and the existing level of manual control.

CONTROL STRATEGY DEVELOPMENT. Degree of Control. The appropriate amount of aeration control that is required or desired, and can be achieved, is site specific for each activated-sludge WWTP. The degree of aeration control implemented can range from the extremes of infrequent manual manipulation based on manual measurements (virtually no control) to comprehensive automated control of the aeration delivery and distribution hardware to maintain the on-line measured DO concentration at desired setpoints throughout the aeration trains under dynamic loading conditions.

NEW WASTEWATER TREATMENT FACILITY. For a new WWTP, the decision to incorporate aeration control is straightforward. The capital cost of implementing even a high degree of automated control, as an incremental cost above that required to provide open-loop on-line monitoring, is a small percentage of the total initial cost of a WWTP, generally 1 to 5% depending on the size of the facility. Successful implementation of automated control of air flow and DO in a new WWTP requires that the plant be designed from the outset with the intent of incorporating efficient automated control. The degree of automation provided should be based on the size of the WWTP and the in-house instrumentation maintenance capabilities.

Careful attention to process and hardware flexibility is necessary to realize the maximum benefits from a well-designed aeration control system throughout the WWTP design life. The various components of the aeration system must be designed to allow for operational flexibility to meet the dynamic oxygen requirements of the system. Of greatest importance in this respect are operational constraints and turndown capacity of the blowers, air distribution control hardware, and diffuser allowable air flow range.

RETROFIT FACILITY. For an existing manually controlled WWTP, the decision to retrofit for automated aeration control must be based on one or both of the following objectives: provision of more effective control of the aeration process to minimize operational problems, or optimization of the aeration process to achieve energy consumption savings.

In either case, the degree of control implemented should be based on an incremental cost–benefit analysis. The economic analysis should consider the cost of retrofitting the WWTP for various levels of control, training requirements and personnel changes, the potential for energy savings, operations and maintenance costs, and the intangible benefits of improved operation and plant reliability. Automated control may also result in more efficient use of existing tanks and optimization of WWTP capacity. This might permit postponement of expected WWTP expansions and the associated capital financing expenditures. While this possibility is a less obvious benefit of improved aeration and process control, it may outweigh the cost savings that are achievable through operating cost reduction alone.

Control Systems. The principles of control theory used in the control systems examined in this section are available in standard control textbooks (Astrom and Wittenmark, 1984; Coughanowr and Koppel, 1965; and Ogata, 1970). In addition, a U.S. Environmental Protection Agency report (Flanagan and Bracken, 1977) describes in detail several control strategies and basic techniques used for DO control in the activated-sludge process.

The automated control of mixed liquor DO concentration to a setpoint value does not itself require any in-depth knowledge of biological metabolism or activity. Simply stated, the DO concentration is measured using on-line instrumentation, and the air flow rate is controlled in the zone of the basin where the measurement is being made. The manipulated variable in this case is adjusted, if necessary, by a controller directing final control elements accordingly.

CONTROL THEORY. The most common DO control strategies involve either conventional feedback or feedforward–feedback controllers. If feedforward action is added to the controller, as illustrated in Figure 9.2, an incoming process disturbance (a change in influent flow rate, for example) is detected directly through additional instrumentation. The controller initi-

---

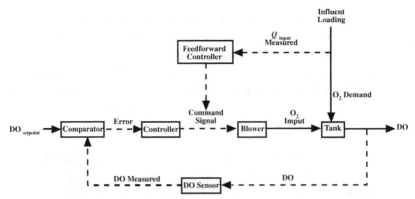

Figure 9.2 **Block diagram for feedforward–feedback control loop (DO = dissolved oxygen, and *Q* = flow) (from Boyle, W.C., *et al.* [1990] *Fine Pore Aeration for Wastewater Treatment.* Pollution Technology Review No. 191, Noyes Publications, Park Ridge, N.J., with permission)**

ates control of the manipulable variable immediately rather than waiting until the disturbance is reflected in a change in the controlled variable. Typically, the two controllers are coupled so that the feedforward portion of the controller paces the action and the feedback portion serves to "fine-tune" or "trim" the feedforward control action. The process DO response to changing loading conditions is generally slow and can typically be handled by control systems using feedback control only.

CONTROL FUNCTIONS. The control signal is derived from the measured variable error according to a control law or algorithm. This can be as simple as on–off control of the final control element to adjust the manipulable variable. For DO control, this could involve bringing on line or taking off line one or more blowers in a multiple-blower system to change the aeration rate whenever the DO is below or above the setpoint. For cases in which the resulting response of the process DO concentration to this type of air delivery control is fast, excessive starting and stopping of blowers can result. This is undesirable from the standpoint of hardware operation and of maintenance and energy demand and consumption.

These problems can be minimized by modifying the controller to take no control action within a user-selectable band about the setpoint. For example, when the desired DO concentration is 1.5 mg/L, the controller could be set to start up a blower at 1.0 mg/L and to shut off the blower at 2.0 mg/L. Thus, the DO concentration would be allowed to fluctuate within a "band" about the setpoint. This type of controller is called a *deadband* controller and results in a longer cycle time for blower starting and stopping. There is, however, a corresponding decrease in the degree of control.

Typically, however, the control signal is derived from a continuous controller such as a proportional-integral controller. This type of control has worked well for aeration control.

Generally, for a given set of operating conditions, there is an optimum set of controller parameters. However, for DO concentration control, in which the dynamics are nonlinear and time-varying, the tuning parameters that are optimum for maintaining control under peak diurnal loading periods may not sustain that level of control over an extended operating range. A compromise between ideal control and the maintenance of controller stability often results in the controller being "detuned" to a degree to achieve adequate control over a wide range of operating conditions. The controller is then periodically retuned, as necessary, to accommodate gross changes in DO dynamics that can occur on a seasonal basis.

Dissolved Oxygen Control Strategies. *CONVENTIONAL CONTROL.* In aeration basins designed to be completely mixed, the oxygen demand is relatively uniform. Therefore, automated control of DO concentration in the basin is based only on feedback from the DO sensor. In systems such as plug flow or step feed, however, multiple control loops may be required to control DO concentration effectively.

The spatially varying oxygen demand along a plug-flow reactor requires a nonuniform rate of oxygen transfer to accomplish uniform DO control. For a steady-state condition, this can be achieved by tapering the diffuser density in grids down the length of the basin. Automated air distribution control valves can be installed to regulate the air flow rate to, and maintain the DO concentration in, each grid at a desired setpoint. If this is not practical, the air distribution profile can be established with manually adjusted air distribution valves, and the total air flow to the basin can be regulated automatically to maintain the desired DO profile down the length of the basin.

Figure 9.3 shows a typical DO control system for a tapered diffused air delivery, compartmentalized, plug-flow aeration basin. A proportional-integral controller determines the required change in air flow needed to restore the DO setpoint and "cascades" a setpoint for air flow rate to a separate air flow control loop. Because the control loops for each basin operate independently, an increase in air flow to basin 1 will result in decreased air flow to basin 2 and decreased DO concentration in basin 2. The system will attempt to compensate for this at the next control interval. This can result in the controllers continually "hunting" and perhaps even cause instability in the control system.

The addition of a pressure control loop minimizes "hunting." This loop ensures the existence of adequate air flow for both basins. The system operating pressure setpoint may be minimized to achieve maximum energy savings by always maintaining one of the valves in its most-open position

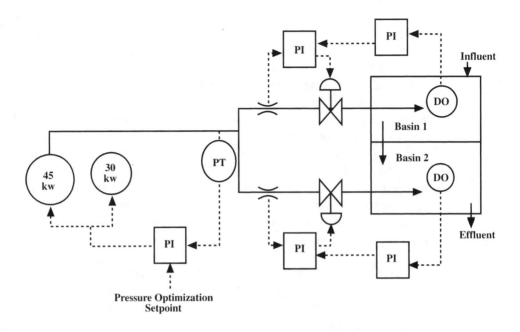

Figure 9.3   Two-stage dissolved oxygen control system for a compartmentalized plug-flow aeration train (DO = dissolved oxygen; PI = proportional-integral controller; PT = pressure transmitter; ✕ = air flow measurement transmitter; and ⋈ = control valve)

or by defining the pressure setpoint and deadband as a function of the total air flow to the system. The control algorithm typically incorporates a minimum allowable air flow setpoint to maintain minimum mixing requirements in the reactor and prevent throttling to less than the minimum capacity of the blower.

For digital controllers, the timing of the control loop action (control interval) is critical. It must be matched to the time constants of the variables being controlled to accommodate process dynamics, process dead time, if any, and disturbance or noise in the process or primary measurement signal that should be ignored or rejected. For the case above, the DO controller is the slowest loop in the system. This is because the aeration process dynamics are relatively slow, and it can take up to 30 minutes for the DO to reach a new equilibrium value after a change in air flow has been initiated, even under steady-state conditions. There is little point, therefore, in running the DO control loop more often than once every few minutes.

The individual air distribution control loops have a much faster response time, and the control interval may be on the order of several seconds. Adequate deadbands must be provided to protect the life of the motor. A pressure optimization control loop will adjust the final control elements on the blowers to regulate the total air flow to the system and maintain the desired system pressure. This controller must account for the disturbances generated by the control valve. The control interval will therefore be slower (approximately 1 to 2 minutes) than the air distribution control loops, but not as slow as the DO control loop. The proper tuning of these several control loops is necessary to obtain the desired system response. This task can become tedious because of the various process and controller interactions involved.

ADVANCED CONTROL. Programmable digital controllers have much more flexibility than traditional analog controllers. These controllers can readily compensate for the considerable dead time that typically delays the response of the control variable to disturbance or control action. They also facilitate the implementation of more advanced controllers that can extend the empirical engineering solutions exemplified by proportional-integral control. Digital filtering of raw signals from process instruments can eliminate or minimize controller disruption resulting from signal noise.

Because the digital controllers to which we refer are actually software packages supplied by a vendor, the degree of sophistication and flexibility will vary. Proper application will involve knowledge of the vendor's algorithm, proper setting of tuning constants, and determination of controller execution rate and deadbands.

Advances in industrial process control have led to the development and commercial application of adaptive controllers, otherwise known as *self-tuning* or *auto-tuning* controllers. With these controllers, the controller tuning parameters can be calculated from measurements of process response to control action and updated periodically as process conditions and DO dynamics change. The two interacting loops of a self-tuning controller are shown schematically in Figure 9.4. The updated tuning parameters of the conventional controller are obtained either directly or determined from process parameters that have been estimated periodically from the process

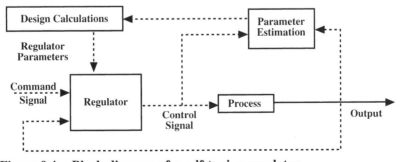

Figure 9.4   Block diagram of a self-tuning regulator

response (Astrom and Wittenmark, 1984). This minimizes the effects of time-varying disturbances on the controller.

The adaptive controller can be implemented for feedforward–feedback situations and also when the sequential changes in the manipulated variable must be constrained. This serves to minimize wear resulting from excessive control action on mechanical components such as motor drives and valves. Self-tuning or adaptive control of DO concentration has been successfully implemented at the Kappala WWTP in Stockholm (Olsson *et al.*, 1985).

CONTROL SYSTEM COMPONENTS. Instrumentation. The successful operation of installed on-line instrumentation is critical to the successful application of automated aeration control. Incorrectly applied, installed, or maintained instruments can render an otherwise well-designed process control system useless. Not only will the objectives of the control system fail to be met, but serious and costly operational problems can result. Therefore, sensors are often considered the weakest links in the control system.

The control strategies discussed in the Dissolved Oxygen Control Strategies section require DO and, in some cases, air flow rate instrumentation. Other process instrumentation requirements for this application generally include air pressure, air temperature, and air distribution valve actuator position monitors. Instrumentation required to monitor and control blower operation depends on the specific requirements of the blower control system. Functions monitored may include speed, blower inlet vane or valve position, current or power draw, and suction and discharge temperature and pressure.

The above instruments are all, with the possible exception of DO monitors, considered standards in the chemical and industrial process control industry. A large body of literature exists concerning the application, installation, calibration, and maintenance of this instrumentation. Some of these references (Considine, 1985; Liptak and Venczel, 1982; Manross, 1983; and WEF, 1993) are considered standards in the industry, and this information should be reviewed in detail before specifying or procuring any of this instrumentation. It is essential that the definitions used to describe instrument characteristics and performance—such as range, span, accuracy, hysteresis, repeatability, sensitivity, deadband, resolution, response time, and drift—be understood to specify and apply these instruments appropriately. Considerations crucial to successful application of the major groups of instrumentation are reviewed briefly below.

DISSOLVED OXYGEN. Typically, DO monitoring equipment is the most troublesome of on-line sensors used in wastewater treatment. It is most often cited as the main reason for DO control system failures. While system failure may result from faulty equipment in some cases, it is at least as likely to result from improper application, poor installation, lack of attention and maintenance by WWTP personnel, or a combination of these. These problems typically have at their root a lack of understanding of the instrumentation, insufficient attention to details during the design phase, or lack of commitment to keeping instrumentation operating properly.

Principles of Operation. Virtually all DO probes available today are electro-chemical cells that contact the fluid through an oxygen-permeable mem-brane. The oxygen concentration in the cell electrolyte reaches equilibrium with that in the bulk fluid, and a chemical reaction induces a change in volt-age across the electrodes. The subsequent current flow across the electrodes produces an electrical signal in proportion to the oxygen content of the fluid. One device has no membrane, and the probe uses the bulk fluid as its electrolyte.

Selection and Installation. Instrument selection should only be undertaken after application constraints such as environmental conditions, operating ranges, and design requirements have been identified. Existing comparative instrument test data and the experience of other users under both bench and field conditions should be taken into account (Kulin *et al.*, 1983, and NCASI, 1984). The user-based, nonprofit North American Water and Waste-water Instrument Testing Association (ITA) conducts structured evaluations in accordance with strict peer-reviewed test protocols. In 1988, ITA com-pleted a comparative test of seven DO measurement systems (North Ameri-can, 1988). Results may be purchased from ITA.

A field protocol for selecting, locating, and maintaining DO sensors has been developed, and based on a 60-day test of equipment from several man-ufacturers, specific recommendations regarding probe location, calibration, and maintenance have been offered (Kulin *et al.*, 1983). The manufacturer typically stipulates installation conditions; however, in general, DO sensors should be installed in the aeration basin in a "dip" mode, that is, directly in contact with the mixed liquor and easily removed and serviced by opera-tions personnel. This minimizes time delays and the maintenance of ancil-lary equipment associated with flow-through cell configurations. The mount-ing hardware should be accessible and of a quick-release type to allow for maintenance (Kulin *et al.*, 1983). Procedures for changing the probe mem-brane and replacing the electrolyte should be easily implemented. It is also important to provide flexibility in mounting hardware and signal cabling to permit relocation of the sensor in the basin (Flanagan and Bracken, 1977).

Field Verification, Calibration, and Maintenance. Each manufacturer has a specific recommended calibration procedure. Typically, it is a simple one- or two-point calibration. As part of the field verification procedure for the installed instrument, accurate output should be verified over the entire expected operating range. Other performance checks, such as response time, hysteresis, and repeatability, should also be made.

In the environment of the aeration tank, the tendency for fouling of the membrane is of substantial concern because fouling affects membrane permeability and, hence, the accuracy of the DO measurement. Fouling in the form of a slimy biofilm is typically the most significant problem. Biofilm fouling generally leads to a low DO reading and thus overaeration. Biofilm fouling can typically be effectively removed from the membrane surface by careful wiping with a wet tissue or soaking in a 10% HCl solution, thereby restoring probe performance. Mineral deposits or oil can change the permeability of the membrane and are not easily removed by cleaning. Manufacturers' cleaning recommendations should be consulted.

Frequency of cleaning will vary depending on process loading characteristics and operating configuration. Generally, a high-rate (low sludge age) system will require more frequent cleaning. To minimize maintenance requirements, it is important to clean the probes only when necessary. Checking the process probe for conformance with a reference probe can indicate when probe servicing is required. The reference probe must be accurately calibrated and have a time response similar to the process probe, and the linearity of both meters must be known.

The frequency for carrying out the conformance procedure can be optimized with experience. Generally, the process probe should not be touched until successive conformance checks show a significant deviation—often taken to be 0.4 to 0.5 mg/L. The criteria ultimately selected is site specific. Cleaning will typically restore probe performance. If it does not, intensive recalibration or even replacement of the electrochemical cell may be necessary. Total annual labor requirements will vary, but 20 to 30 minutes per probe per week should be allowed for conformance checks, cleanings, and calibration when a low level of fouling is encountered.

AIR FLOW MEASUREMENT. Air flow monitoring equipment is an important component of most aeration control schemes. Effective distribution of air flow to several points in the aeration basin requires accurate air flow measurement. Air flow measurement is necessary to ensure that a minimum rate is maintained for good mixing and good distribution among diffusers. Also, the air flow can provide a check on DO probe performance in that a much higher or lower air rate than normal may indicate probe problems. The importance of this measurement demands that the required instrumentation be carefully selected and installed to maximize the probability that performance expectations will be met.

Principles of Operation. The various air flow metering systems widely used today are based either on differential pressure across a control element or mass flow. Differential pressure meters all use the same basic relationship to measure flow rate:

$$\text{Flow rate} = (\text{Velocity})(\text{Area})(\text{Factor}) \tag{9.1}$$

The differences between differential flow meters are largely a function of how the velocity term is determined. Plate orifice and Venturi meters use a constriction to produce a measurable differential head or pressure. This differential pressure is then converted to velocity using fundamentals of mass and energy conservation. To obtain a measurement of mass flow, temperature and pressure corrections must be applied.

Mass flow meters generally operate on the principle of a hot wire anemometer. A wire is placed in the flow stream with an electric current applied to maintain the wire at a preset temperature. The rate of cooling of the wire, based on the current required to maintain it at the preset temperature, is proportional to the mass flow rate of air.

Selection and Installation. Numerous air flow instruments are available, each with its own advantages and disadvantages. These are discussed in detail in the references previously cited (Considine, 1985; Liptak and Venczel, 1982; Manross, 1983; and WEF, 1993). Care must be taken in selecting the air flow meter so that the air delivery system is not unnecessarily constrained in terms of head loss or turndown requirements. It is also important that the meter be sized to accommodate only the future expected ranges in air flow that are expected at that time. Oversizing and attendant poor flow measurement will occur if the range used to size the meter is based too far into the future. Accurate information concerning diurnal variation in aeration demand may not be known if DO control has not been previously employed. Estimates may be made, if possible, from known flow and organic concentrations at the WWTP.

Careful attention to design and installation is needed to avoid jeopardizing the performance of an appropriately applied and sized meter. This includes closely following the manufacturer's recommendations for approach and downstream conditions to ensure the accuracy of the velocity determination.

Field Verification, Calibration, and Maintenance. The fundamental principles outlined for commissioning and maintaining DO sensors also apply to air flow monitoring equipment. The meter typically comes calibrated to the user's specification from the factory. Once the meter is installed in the system, a conformance check should be made, if possible. Even an approximation of conformance using blower performance curves is worthwhile and can likely be done for the entire range of the meter (Speirs and Hill, 1987).

Because the sensors are not in contact with wastewater, much less maintenance is required, once their performance is verified, than for DO probes. Periodic conformance checks are recommended, however.

PRESSURE AND TEMPERATURE. Pressure and temperature measurements are used in the aeration control system to monitor blower suction and discharge conditions. They also provide on-line information for converting volumetric field flow rates to standard flow rates. This is necessary if seasonal air usage comparisons are to be made, but may not be necessary for proper DO control. The seasonal variation in discharge air temperature from a blower, even in cold climates, may not significantly (less than 5%) affect the standard flow rate, provided the meter is calibrated to read out in standard cubic feet (cubic feet $\times$ [2.832×10^{-2}] = cubic metres) at some midrange value of temperature and pressure. The actual air flow is not important if the DO requirements of the tankage are being met.

These instruments are standard throughout the process control industry and are not discussed in detail here. The same fundamentals described for DO and air flow rate measurements concerning selection, verification, calibration, and maintenance apply to these instruments. It is most important to incorporate instruments that are suitable for the environment encountered at a WWTP and are able to meet the objectives of the application in terms of operating range and performance.

Blowers and Air Distribution. The final control elements in any control system are required to carry out the desired control action. For a DO control system, this means adjusting the delivery and distribution of air through manipulation of blowers and air control valves or mechanical surface aerators.

AIR DELIVERY BLOWERS. The two major classifications of blowers typically used in aeration control systems are rotary lobe positive displacement (PD) blowers and centrifugal blowers. The effective and efficient control of the blower or blowers in an aeration control system is almost totally dependent on good design. The blowers must be sized such that an operating map of the air delivery system not only meets the expected variations in process air requirements, but also maximizes blower efficiency throughout the delivery range. The control system must be designed not only to minimize unnecessary disturbances in air flow rate and DO concentration, but also to maximize the energy savings extracted from variable air delivery. The control strategies for the two different classes of blowers are different because of their principles of operation.

Figure 9.5 illustrates the general operating characteristics of a rotary PD unit and a centrifugal unit. As seen in the figure, the rotary PD unit will deliver a relatively constant air flow rate over a range of discharge pres-

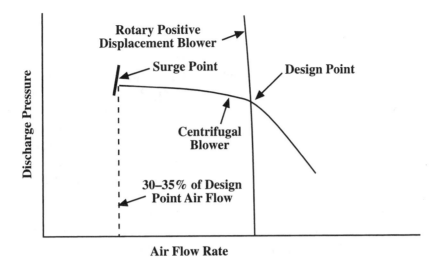

Figure 9.5   General operating characteristics of blowers

sures. In contrast, the centrifugal unit is capable of delivering a range of air flow rates at a relatively constant discharge pressure.

High efficiency and the ability to operate over a range of discharge pressures are two of the principal advantages of PD blowers. The main disadvantages of these units, compared with centrifugal blowers, are the inability to effectively throttle air flow rate (however, the blowers can be speed-controlled), the usual requirement for a more substantial foundation to dampen and resist vibration, and generally noisier operation.

Advantages of centrifugal blowers include quieter operation and smaller foundation requirements. Their disadvantages include a limited operating pressure range and a reduced volume of air delivered with any backpressure buildup as a result of clogged diffusers.

Turndown Considerations. If the selected blowers are to operate in an economic range over the entire life of the project, they must be capable of supplying appropriate volumes of air to meet the varying oxygen demands of the wastewater. Oxygen demand variations result from both diurnal fluctuations and the differences between start-up and design loads. Therefore, blower selection should take into account minimum air requirements at WWTP start-up (which may be limited by mixing requirements for fine-bubble diffusers) and peak air requirements at design conditions. It is essential that appropriate aeration control strategies and equipment be incorporated in the overall air delivery design. This will ensure full realization of the potential operating benefits of reduced power consumption of fine-pore diffusion systems compared with coarse-bubble and mechanical aeration systems. This aspect of blower design assumes even greater importance in retrofit situations.

Positive Displacement Blowers. Positive displacement blowers are essentially constant-volume, variable-pressure machines. The pressure at which the blower runs is dictated by system requirements. Historically, changing the number of blowers in service and controlling blowers with multiple-speed motors were the only options for controlling PD blowers. Today, use of a variable-frequency alternating current (ac) drive also allows the PD blower to run as a variable-volume, variable-pressure machine. Controlling and saving air flow, however, must translate into energy savings. Variable-frequency drives can consume up to 10% of the energy applied as heat. This premium must be considered carefully because it can sometimes affect the economic incentive for automated DO control.

From an operations viewpoint, the variable-frequency drive should not be permitted to lower the blower speed to less than the manufacturer's recommended limit. Less heat dissipation at lower-than-recommended speed could result in overheating of and damage to the blower. At the other end of the speed scale, most ac variable-frequency drives can operate at 110% of normal frequency, thereby increasing motor speed and, hence, blower speed by 10% overall. While this places a greater load on the motor and consumes additional energy, it is one way to increase the capacity and operational flexibility of the PD blower.

Another alternative is to drive the PD blower with a gas engine at facilities where methane gas is a byproduct of the anaerobic digestion process. Automatic control of the engine fuel governor may be provided to control the speed of the PD blower. The engine efficiency will stay relatively high over the range of acceptable blower speeds.

If speed control is not employed for a PD blower, then caution must be exercised in throttling valves on the distribution system. Throttling will raise the system pressure and redistribute air flow. Most PD blowers have a maximum safe operating limit. Even if speed control is used and throttling of air valves is employed, it is wise to have some sort of fail-open algorithm for the air valves in case of blower speed control failure.

It is important that the blower and drive manufacturers understand in detail the application, control strategy, and anticipated operating conditions of the air delivery system. This ensures proper integration of appropriately sized motors, drives, and monitoring instrumentation.

Centrifugal Blowers. Typically, the volume output from centrifugal blowers is manipulated by adjusting speed or inlet guide vanes or by throttling the inlet valves. These approaches vary in terms of difficulty, energy efficiency, reliability, and effect of stability. Overall blower efficiency will drop significantly as the inlet guide vane or valve is throttled. The control of these blowers is also complicated by the need to stay above the low-output operating limit or surge point. Often, large, complex blower packages come with built-in controls to prevent surge conditions. If built-in controls are not provided, the digital controller should have feedback of either amperage or power use to limit operation of the guide vane or butterfly valve position. Limiting the movement in this manner may be more successful than built-in controls because the limits may be more easily integrated with the operational control strategy. The operational control strategy may use position, amperage, or power as the input and output for controlling the position of the throttling guide vane or butterfly valve on the blower. Detailed explanations of various types of centrifugal blower control schemes are available elsewhere (Flanagan and Bracken, 1977; Lutman and Skrentner, 1987; Nisenfeld, 1982; and WPCF, 1988).

The same degree of involvement of the manufacturer with the designer and instrumentation and control specialist as is needed for a PD blower application is essential for centrifugal blowers.

Multiple Blowers. Manufacturer blower curves should be used to develop a detailed operating map for configuring the number of blowers in service and their individual or collective operating points to achieve the most efficient air delivery possible at all times. This map should incorporate the effects of environmental conditions such as temperature and humidity. Once the blowers are installed, it is possible to fine-tune this operating map and determine any site-specific operating limitations through measurements of temperature, humidity, pressure rise, power draw, speed, and overall air flow.

Mathematical and empirical relationships incorporating these parameters can be used to optimize the on-line configuring of the air delivery system. Information should be obtained from the blower manufacturer application engineer to confirm the most efficient operating configurations to meet any operating setpoint. Any operating limits or restrictions identified by the application engineer should be strictly observed.

Start–Stop Control. For any aeration system, careful attention must be given to blower start–stop control. A well-designed automated control system is able to use the maximum range of a blower, or combination of blowers, to minimize the necessity of bringing on line or taking off line additional blowers. Implementation of automatic start–stop of blowers is straightforward. In some control systems, however, operator approval is a required additional judgment step prior to start-up or nonemergency shutdown of any blower to minimize starting and stopping of blowers. In other cases, the control system only flags the need for bringing additional blowers on line or taking some off line, and the actual operation is carried out manually.

To minimize operational problems, a blower (whether automatically or manually initiated and controlled) is typically started off line and brought up to operating conditions by delivering air through either a recycle loop or vent valve. Only then are the appropriate isolation valves manipulated to allow the blower to discharge into the common header and be integrated into the control loop. Under certain operating and start-up conditions, the main header operating pressure may have to be reduced by overrides in the aeration control valve strategies to permit the blower to come on line. In addition, to minimize energy demand charges, it is often necessary to reduce air delivery, and thereby power consumption, of the running blowers before bringing a new blower on line. Finally, the control strategy must respect any restrictions on minimum run time and time between starts.

The above considerations are machine and facility specific. For the design example, specific operating requirements for the selected blowers would be incorporated in the initial control software design. They would be fine-tuned following system start-up and after a modest level of operating experience had been obtained.

Variable-Capacity Control for Centrifugal Blowers. For centrifugal blowers designed to operate at relatively constant system pressure, there is a nonlinear relationship between inlet valve or guide vane position and blower throughput. However, blower power draw or motor amperage versus throughput is nearly a linear relationship and may be used to control the inlet guide vanes or valve as shown in Figure 9.6.

When multiple air valves are manipulated by independent automatic DO controllers, generally in more complex control systems, a main header pres-

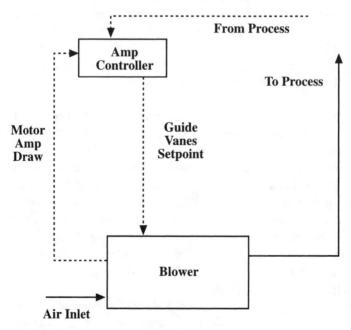

Figure 9.6   Blower inlet guide vane control schematic

sure controller is added to minimize the disturbance of air distribution control on the blower controller. The pressure controller cascades the amperage or power draw setpoint to the amperage controller as shown in Figure 9.7. The addition of a variable-pressure setpoint optimization routine may be justified in this case to further maximize aeration energy savings by always minimizing the header operating pressure. However, because of the relatively small size of the blowers in this example, this refinement would most likely be justifiable only after the design capacity of the WWTP was being approached and operation of three blowers was required to meet the normal oxygen demand.

Surge Protection. When a centrifugal blower cannot develop enough pressure to overcome the downstream process pressure, the blower can begin to surge—a condition in which momentary reversible pulsing of air occurs inside the compressor. Because this condition can be particularly destructive to centrifugal blowers, surge protection is essential. Most blower manufacturers provide independent surge protection control systems; however, additional levels of surge protection can, and should, be incorporated in the overall blower control algorithm. Surge can occur at certain high-pressure or low-throughput conditions. Therefore, careful attention to the interaction between air distribution and air delivery controllers is necessary to minimize line pressure.

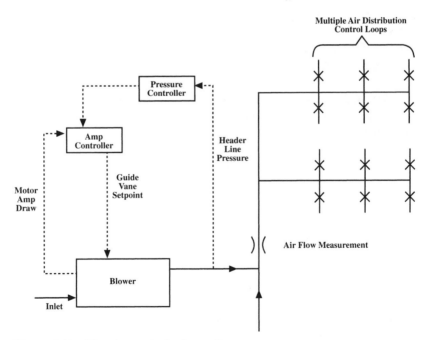

Figure 9.7   Blower control schematic

A minimum blower amperage limitation that corresponds to the above-mentioned critical operating conditions is often used for surge protection. However, when inlet guide vanes are manipulated to control centrifugal blower delivery, as in this example, the surge point varies with each guide vane setting. Typically, when guide vanes are manipulated to reduce blower throughput, the motor amperage draw corresponding to the surge point is also reduced. Care must be taken, therefore, not to set the absolute minimum amperage too high, thereby reducing effective blower turndown under normal operating conditions. The relationship between guide vane position and surge point amperage draw is often nonlinear over the range of guide vane operation. Once established, it can be used with on-line measurements of blower differential pressure or individual blower discharge air flow to predict and appropriately update safe operating limits.

Blower Performance Evaluation. The initial design should incorporate provisions to evaluate periodically the performance of the blower control system. This is particularly important when multiple blowers operate in parallel. At various periods in the WWTP design life, the operating objective may be either to balance the load among blowers or to maximize air delivery overall efficiency in accordance with the operating map. To confirm that these objectives are being achieved, provisions must be made for measuring individual blower operating parameters. Appropriately located standard pipe taps must be included in the suction and discharge piping of individual blowers to facilitate temporary installation of instrumentation for measuring such variables as temperature, pressure, and air flow rate. These data provide a means of assessing individual machine and overall system performance.

AIR DISTRIBUTION. Control valves are used to distribute air to aeration basins, headers, and grids. Control valves are designed to maintain a relationship between air flow rate through the valve and valve travel as it is varied from 0 to 100% of its most-open position. Valves are thus characterized as providing relatively uniform control loop stability over the expected operating range.

A valve with a linear flow characteristic yields an air flow rate that is directly proportional to valve travel. For an equal-percentage valve, equal increments of valve travel produce equal percentage changes in the existing flow. Equal-percentage valves are typically used for pressure control applications, in which a large percentage of the pressure drop is typically absorbed by the system itself, with only a small amount available at the valve.

More often, standard butterfly valves will be used for air distribution. When more than 50% open, these valves may effect little change in air flow rate for a given valve movement compared to the change for an equal movement when they are less than 50% open. To prevent valve oscillation problems, tuning should be made with conditions of high header air pressure and valve position less than 50% open. Control, however, will not be as responsive when the air pressure is low and the valve is more than 50% open. Care must be exercised in design so that the valve size is not so large as to provide unresponsive action over too large of a range, for example 25 to 100% open.

The sizing of valves, not unlike air flow meters, is critical for good air distribution control. The valve should be sized for control over the immediately foreseeable operating range. In some cases, valves with replaceable "trim" are available, which provides the flexibility to handle different ranges of flows.

A valve and actuator may have a degree of hysteresis associated with their response to a control signal. Hysteresis is a measure of the difference in the valve response for a particular input signal, depending on whether the new position was approached from a more-open or more-closed position. If such a condition exists, it must be accommodated in the control algorithm to obtain good control.

Valve motors may be controlled either by an analog output or by a series of digital pulses from the controller. If an analog output is used, then the controller or valve motor must have a mechanism for holding the last valve position in case of failure of the signal from the controller. With a motor using digital pulses, this is not a problem because the controller will be unable to generate pulses to open or close upon failure. A controller generating digital pulses may be capable of generating a signal of such short duration that the motor contact will close, but no actual valve movement will occur. This will contribute to shortening valve motor life even though no valve control action has been taken. Adequate deadbands must be provided to ensure that this will not happen. This is typically not a problem in motors taking an analog signal because the valve motor controller will have an internal comparator that will prevent movement if the difference between the position signal and the output signal is too small.

Valve motors may also be either ac or direct current (dc) motors. An ac motor may have a longer service life, but the speed of travel of a dc motor is adjustable, while that of an ac motor is not. Consideration should also be given to availability of replacement parts or motors during the selection phase.

It is also important to note that, while the valves should always be maintained as open as possible to minimize overall system pressure and aeration energy expenditure, it is necessary to sacrifice some head loss to have control. This tradeoff requires constant balancing. Energy losses resulting from increased head loss can be minimized with properly sized and applied valves. The air distribution designer should work with the valve supplier applications engineer in selecting and specifying these valves and their operating characteristics.

Air Control System Default. It is important that the control system be designed to permit override or suspension of automated control and allow manual operation. Further, the control system should default to a safe operating condition that maintains process integrity in the event of control system failure.

For example, the blower control system could fail, resulting in excessive pressure to the aeration tanks. The air control valves could start throttling to reduce air flow to the required setpoint, eventually resulting in the surge of a centrifugal blower or a high pressure that could damage a PD blower. In this case, either a high-pressure alarm or low-position alarm could cause the air valves to travel to their most-open positions and prevent further problems. Another example could be blower rotation. The total system air flow and pressure could vary significantly during rotation of blowers, causing excessive valve movement and possibly problems with blower surge. Programming should include a way to bypass valve control during this period, possibly bringing the valves to their most-open positions and leaving them there until rotation is complete.

AERATION CONTROL EXAMPLE. The general layout of process air piping for an example is presented in Figure 9.8. For this design, the diffusers in each plug-flow reactor are arranged in three grids or zones to allow for adjustment of air flow down the length of the reactor to more closely match the anticipated spatial oxygen demands. As stated previously,

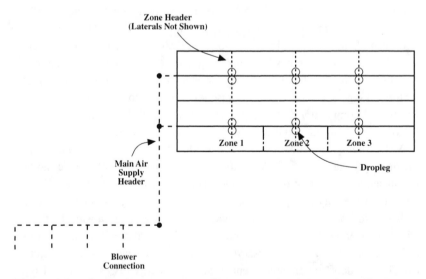

Figure 9.8   General arrangement of process air piping (from Boyle, W.C., *et al.* **[1990]** *Fine Pore Aeration for Wastewater Treatment.* **Pollution Technology Review No. 191, Noyes Publications, Park Ridge, N.J., with permission)**

the degree of aeration control implemented can generally range from the extremes of infrequent manual manipulation, based on manual measurements, to comprehensive, automated, setpoint DO concentration control. These options also exist for this example.

Ideally, a setpoint DO concentration would be maintained under dynamic loading conditions in each of the 12 aeration zones shown in Figure 9.8 by automatically configuring the blower system to efficiently meet total air requirements and by manipulating the air distribution control valves to vary the air delivery rate to each aeration zone. This would result in a fairly complex control system. The least complex control system for this example would be to automatically manipulate the blower configuration and total air output to the four aeration trains based on maintaining a desired DO setpoint at one point in one of the reactors. The desired DO concentration profile along the length of each reactor would be maintained by periodically manually adjusting air delivery to each zone.

For this example design, several control options lie between these extremes, each representing a compromise between the level of control achievable and the complexity and cost of the control system. These options are discussed below.

Air Delivery Selection. The basic air delivery control hardware requirements are similar for virtually all anticipated control options. Four 1 320-L/s (2 800-scfm) blowers could accommodate the diurnal, seasonal, and yearly variations in total air requirements expected when the WWTP reaches its design loading. The air delivery control strategy for any option considered would continually optimize the number of blowers in service and their operating points to deliver efficiently the total air required at any time in the design life of the WWTP. This strategy is essential to achieve the goal of maximized energy savings through effective air flow control. Often, a well-designed control system maintains DO concentration setpoints as desired but, because of inefficient operation of the blowers, does not realize its full potential aeration energy savings.

Air Distribution Selection. Control valves are required for effective distribution of air to the individual zones in each aeration train and, depending on the control system, to the aeration trains themselves. Whether automatically or manually adjusted, the valves must be properly sized and of a type suitable to minimize pressure head loss yet maintain controllability over their anticipated ranges of air flow. The manufacturer applications engineer should be consulted to ensure that type, size, and flow control characteristics are compatible with control objectives over the anticipated design life of the WWTP. For automated control, electric or pneumatic valve actuators, complete with positioners when necessary, are used in place of manual hand operators.

To illustrate how the design of the air distribution control system can be incorporated into the overall aeration system design (see Figure 9.6 for air delivery system layout), two control options are outlined below for the design example. The first represents a low-complexity option and the second a moderate-complexity control option. A third control scheme, not recommended for this size WWTP, is also presented to show a high-complexity control strategy that could be considered for large plants. These options are presented to illustrate the various degrees to which aeration control can be implemented. The specific features of any option could be integrated to generate a modified—that is, a more- or less-complex—control scheme, as appropriate.

Strategy Development Options. *LOW-COMPLEXITY CONTROL STRAT-EGY.* The least complex control strategy for this example involves manipulating the blower configuration and total aeration output to the four reactors to maintain a desired setpoint DO concentration at one location in one aeration basin. Initially, zone 2 is selected for the DO probe location for this control measurement. Zone 3 is not selected because load changes and changes in system oxygen demand may not always be detected in this zone or will be detected too late to respond to changes in DO concentration in zones 1 and 2. Zone 1 is not selected because DO changes typically occur more rapidly here than further down the tank and may result in unnecessary or erroneous corrections to air flow in zones 2 and 3. This is discussed further in the section below titled Dissolved Oxygen Probe Location and Dissolved Oxygen Setpoint.

A proportional-integral DO controller cascades the air flow setpoint to the air demand controller, which regulates the blower output through the manipulation of inlet guide vanes to maintain an amperage setpoint, as discussed above in the Air Delivery Blowers subsection, Variable-Capacity Control for Centrifugal Blowers. The operator would strive to maintain an acceptable DO concentration profile along the length of the reactors by periodically manipulating the aeration grid distribution valves manually to adjust air delivery to other aeration zones. In this option, careful operator attention is required to maintain the distribution valves in their collective most-open positions to minimize air header pressure.

The control loops for this option are shown schematically in Figure 9.9. The blowers are operated by the control system to respond to variable oxygen demands. The number of on-line blowers depends on the load to the WWTP. Bringing blowers on line or taking them off line is carried out automatically upon receiving an on–off signal from the air demand controller.

The blower system operating map would be used to determine control of the on-line blowers to achieve optimum energy efficiency. This may be accomplished most effectively by controlling all on-line blowers with the

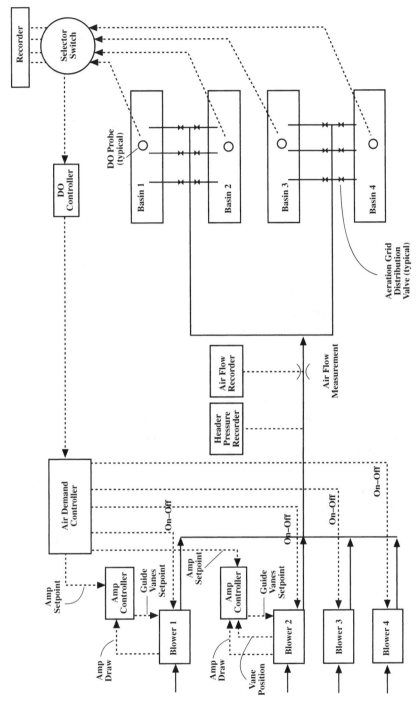

Figure 9.9   Low-complexity control schematic (DO = dissolved oxygen)

same signal from the air demand controller. This strategy controls all on-line blowers at the same operating point while matching the variable total air flow demand.

Alternatively, one blower could be operated by the control system to respond to variable oxygen demands and one or more of the other blowers operated at constant output to provide a base supply of air. The base supply could be provided by a PD blower that would operate efficiently over an acceptable range of discharge pressures. Periodic substitution of a different blower to serve as the variable delivery blower allows for load balancing and accommodates maintenance requirements. The blower manufacturer application engineer should be consulted to ensure that the envisioned control strategy achieves the control objectives.

It may be desirable initially to override part of the automated control of the blowers and allow the operating staff to control manually certain aspects of the blower system operation. This could be accomplished by providing the controller output information to the operator, who would in turn make the required adjustments to guide vanes and to the bringing on line or taking off line of blowers. As the level of comfort with the control system increases, fully automated control could be implemented on a staged basis.

This control strategy assumes the four parallel aeration basins are operated to achieve uniform spatial and temporal oxygen demand profiles and that by similarly adjusting the air distribution valves in the parallel zones, an acceptable DO profile can be obtained in each aeration basin. However, even though the aeration basins are operated in parallel, it is unlikely they will perform identically.

An important assumption inherent in this strategy is that the system hydraulic design results in equal wastewater flow distribution and that air system head losses are similar, if not equal, for each aeration basin. This may not necessarily be the case.

Ideally, an on-line DO sensor would be provided for each aeration zone (a total of 12 sensors) to optimize monitoring of system DO concentration dynamics. However, for reasons outlined in the Control System Components subsection, Instrumentation, it may be appropriate to reduce, at least initially, on-line probes to a number with which operation and maintenance personnel are comfortable.

Dissolved oxygen monitoring can be accomplished by placing DO probes with a continuous readout in the other three basins at the same location as the DO probe in the control basin. This design permits any of the four readout probes and their respective basins to serve as the DO control system, depending on which basins are out of service for cleaning. This arrangement also allows any of the basins to be removed from service during low loading periods.

A calibrated portable DO probe can be used by the operator to monitor and maintain desired DO levels in those sections of the system not served by on-line meters. As acceptance of on-line instrumentation and the comfort level of the WWTP staff increase, additional on-line probes could be added. It is important that hardware and software provisions be incorporated in the control system design for easy accommodation of expanded monitoring capability.

The tuning of the rather coarse aeration controller recommended for this low-complexity strategy would be carried out as discussed above in the Control Strategy Development subsection, Control Functions. Selection of the final DO feedback location and setpoint would be adjusted following system start-up and after process response was known.

MODERATE-COMPLEXITY CONTROL STRATEGY. A moderate-complexity control strategy is described here as an alternative to the low-complexity control system to provide more exact DO control in each basin. This system also facilitates more accurate control of air flow to each basin by using individual DO setpoints, controllers, air flow control valves, and air headers for each basin.

The basic controller design is shown schematically in Figure 9.10. The major difference between this design and the low-complexity control system is that each aeration basin is provided with its own separately controlled air distribution header. Thus, the air control systems for the four basins are independent of each other, and the need to assume, or dictate, that adjacent aeration basins are operated identically is eliminated. The control of the DO concentration profile in each aeration basin would be similar to that described in the low-complexity example.

One measurement of DO concentration initially in zone 2 of each aeration basin would provide feedback to the air flow controller for that basin. Again, as indicated in the low-complexity example, it would be ideal to monitor the DO concentration in each of the other zones with on-line probes. However, the DO concentration could, initially at least, be monitored with a portable DO probe and meter in the other zones. Periodic manual adjustment of the air distribution valves in the individual aeration zones maintains the desired DO concentration profile. This would be easier to achieve than in the previous case because of the greater independence of the four aeration control system. As in the low-complexity example, the operator would strive to keep the air distribution valves as collectively open as possible to minimize header pressure.

Automated valves located in the four individual headers distribute the total blower output to the four aeration basins. At least one of these valves is always maintained in its most-open position to minimize the main air header pressure. An alternative to using valve positions to achieve this goal would be to set the pressure setpoint as a function of total air flow to the

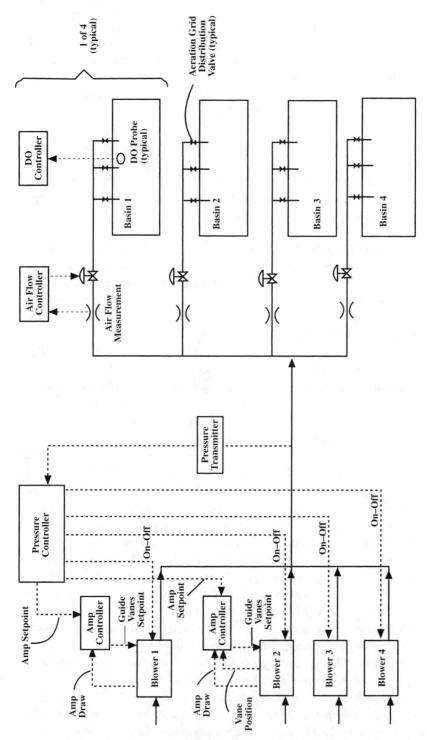

Figure 9.10   Moderate-complexity control schematic (DO = dissolved oxygen)

system. Definition of this function would be based on testing and experience. A pressure controller located in the main header regulates blower output by manipulating the inlet guide vanes as described above. The same considerations regarding blower control options for achieving optimum system efficiency apply to this control strategy as well.

For this strategy, implementation of fully automated air distribution control could also be staged to build operator confidence. Initially, the control strategy would be simplified. The operator would designate the basin with the highest oxygen demand as the "control basin" and use the output from the DO controller in that basin to manually set its air header distribution valve in its most-open position. The DO controllers for the other three basins would automatically adjust their respective air distribution header control valves to maintain the desired DO setpoint in their respective basins.

Implementation of the four air flow controllers shown in Figure 9.10 could be postponed, and the DO controllers could be used to manipulate the air distribution control valves directly. As before, the manual aeration grid distribution valves in each basin would be adjusted by the operator to achieve the desired profile. The output from the DO controller with the manually controlled header valve would be used to control the blowers by providing the setpoint for an air demand controller, as shown in Figure 9.9. Installation of the pressure controller shown in Figure 9.10 would also be postponed. Once experience is gained with the air header distribution control valves, and the operating staff has accepted automated control of these valves, full automated control of the air distribution system can be implemented if desired.

HIGH-COMPLEXITY CONTROL STRATEGY. A more complex controller design option for an aeration system configuration similar to the one used in this example provides independent setpoint control of DO concentration in each zone or grid of each of the four parallel aeration basins. This control strategy would typically be considered only for a much larger WWTP (more than 890 L/s [20 mgd]) because it is probably not cost effective for the plant size used in this design example. This system uses 12 cascaded DO concentration/air flow control loops to control the air to each zone and a main air header pressure controller to regulate blower output through manipulation of inlet guide vanes to maintain an amperage setpoint (as discussed above in the Air Delivery Blowers subsection, Variable-Capacity Control for Centrifugal Blowers) and the number of on-line blowers. The control system is shown schematically in Figure 9.11.

The complexity of this control scheme requires that additional instrumentation and final control elements be provided. In addition to the 12 DO sensors, air flow measurement (standard conditions) is required for each zone. An automated actuator/positioner is also required for each aeration grid dis-

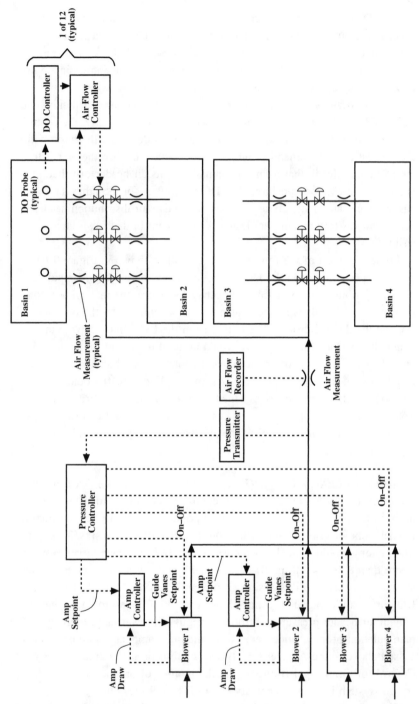

Figure 9.11   High-complexity control schematic (DO = dissolved oxygen)

tribution control valve. These valves must be appropriately selected, sized, and located to achieve the necessary control action. As for the moderate-complexity strategy, a most-open valve algorithm may be incorporated to ensure that whatever the air demand, the system will operate at the minimum possible pressure. An alternative would be to vary the pressure set-point based on total air flow requirements.

In each of the control strategies described above, tuning of control parameters and selection of control loop intervals are critical. The control intervals will be on the order of those discussed above in the Dissolved Oxygen Control Strategies subsection, Conventional Control (for example, 20 minutes for the DO control loop, 30 seconds for the air flow control loop, and 3 minutes for the pressure control loop). Fine-tuning of overall control performance will be required once the process is on line and operating at desired DO concentration setpoints.

Dissolved Oxygen Probe Location and Dissolved Oxygen Setpoint. The selection of DO concentration control points and setpoints is an important consideration in the successful operation of any DO/air flow control system. As discussed previously in this chapter, selection of the operating DO concentration affects process performance and aeration efficiency.

In two of the examples given above, DO concentration is set and controlled at only one point in the plug-flow reactor, and air flow distribution is manually adjusted to maintain an acceptable DO concentration profile. The DO profile is dynamic, changing in response to influent concentration and hydraulic load variations throughout the day.

The DO profile will also depend on the type of diffuser system installed and the way the process is being operated. A fine-bubble diffuser system will typically have low oxygen transfer efficiencies (6 to 8%) near the head end of a plug-flow tank, but high efficiencies (15 to 20%) near the tail end. The transfer efficiencies of a coarse-bubble system vary much less, perhaps from 5% at the head end to 8% at the tail end. If fine-bubble diffuser systems are used in a high-rate system (low sludge age), the oxygen transfer characteristics have been found to more closely approximate those in a coarse-bubble system. Because of the differences in oxygen transfer characteristics with diffuser design and operational parameters, it is best that the DO probe location be kept flexible until experience and DO profiles have defined the best location.

The limitations of having only one DO concentration reading in a plug-flow reactor are illustrated in Figure 9.12 (Olsson and Andrews, 1981). In this example, three different shapes of the profile can exist if DO is measured only at point C. If profile 2 is the desired profile for optimum process performance, profile 1 suggests that the biological reactions have been completed too soon, with a corresponding wastage of air. Profile 1 could also demonstrate the profile achieved with fine-bubble diffusers and a transfer

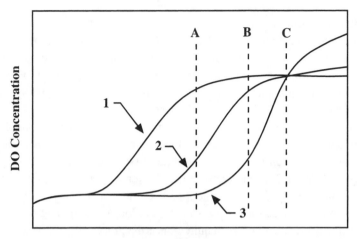

Distance Along Aeration Tank

Figure 9.12 **Significance of additional dissolved oxygen measuring probe on interpolation of dissolved oxygen profile (reprinted from Olsson, G., and Andrews, J.F. (1981) Dissolved Oxygen Control in the Activated Sludge Process. *Water Sci. Technol.,* 13, 341, with permission from Elsevier Science Ltd, The Boulevard, Langford Lane, Kidlington OX5 1GB, UK)**

efficiency that increases quickly after the head end of the tank. Conversely, profile 3 suggests the reactions have not been completed in the desired time, or this could be the profile achieved with coarse-bubble diffusers and a relatively constant transfer efficiency. The use of three on-line DO measurements (that is, point C plus points A and B) provides valuable additional information, particularly near the area of the inflection point in the DO profile.

For profile 1, the early completion of the reactions and corresponding wastage of energy are confirmed by the high value of DO at point A. Point A on profile 2 suggests partial completion of the biological reactions with neither underaeration or overaeration at this location in the basin, while points B and C indicate the reactions have been substantially completed for a long enough period to achieve good process removals and solids settleability. Conversely, the DO readings on profile 3 would seem to indicate either excessive process loading or underaeration, or both, in the critical middle segment of the reactor.

This analysis (Olsson and Andrews, 1981) illustrates the considerations that must be taken into account when trying to maintain an acceptable residual DO profile in a plug-flow reactor. Further, the profiles in Figure 9.12 show that an estimate of the profile slope, based on two or three DO probes placed near the outlet end of the aeration basin, would be a better feedback variable than a measurement from a single DO probe. This could be easily implemented using modern microprocessor-based controllers and the reliable DO monitoring instrumentation available today.

For the high-complexity example shown in Figure 9.11, if 12 probes are provided initially, the three probes in each basin should be used in the early phase of operation to adequately define the location and profile of the DO inflection curve. These data can then be used to adjust process operating parameters as required. The DO setpoint should be high enough to ensure that variations in DO in other parts of the basin (primarily in zone 1), which result from this relatively high-variance control strategy, do not become rate-limiting for extended periods of time. Once the process has been fine-tuned and is performing satisfactorily on a consistent basis, the optimum location of the probes can be determined. Subsequent significant changes in process operating conditions may require reestablishment of optimum probe locations.

In the more complex control strategy, the controller is designed to maintain a DO concentration setpoint in each zone. It is possible that an air flow setpoint would only be used in a zone based on the time of day or on the demand in the other zones. The DO setpoint values are selected to maintain the lowest possible DO concentration residuals without adversely affecting treatment performance. To achieve nitrification, higher setpoints may be required than would be necessary to meet non-nitrifying requirements. In this case, the reduced variance of the controller permits tighter DO setpoints to be maintained without adversely affecting performance. The traditional assumption that DO setpoints should always remain constant may not be valid (Hermanowicz, 1987, and Olsson *et al.*, 1985). Site-specific operating experience is necessary to determine the most appropriate DO setpoint values.

STEP-FEED CONTROL

In the conventional activated-sludge operation, the influent wastewater to the aeration basin enters at the front end of the aeration compartment. The MLSS concentration is relatively constant throughout the basin. For aeration basins with tanks-in-series or plug-flow hydraulics, the MLSS distribution within the aeration basin can be altered to reduce the solids loading to the secondary clarifiers using step-feed operation (Buhr *et al.*, 1984).

During step-feed operation, influent wastewater is added at one or more points along the length of the aeration basins (Figure 9.13). The RAS feedpoint is maintained at the head of the aeration basin. Because the MLSS concentration is less than the RAS suspended solids concentration, the shift in influent feedpoints increases the solids concentration upstream of the feedpoint and decreases the concentration downstream of the feedpoint (Figure 9.14). The net effect is a reduction in the solids concentration fed to the secondary settler for a given flow or mean cell residence time (MCRT). The time to achieve the redistribution of mixed liquor solids is the same order of magnitude as the aeration basin hydraulic retention time.

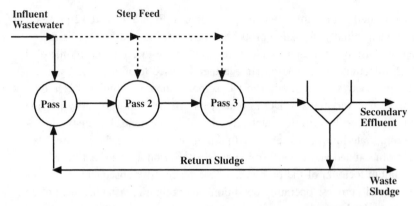

Figure 9.13   Converting the activated-sludge process to step-feed operation

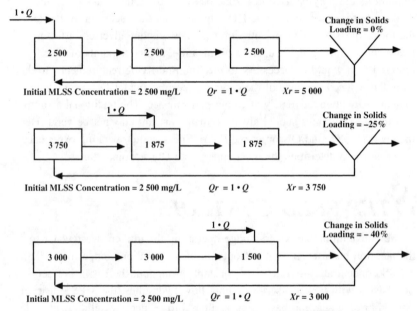

Figure 9.14   Mixed liquor suspended solids (MLSS) distribution during step-feed operation

In extreme step-feed operation, the process is essentially converted to contact stabilization. Substrate is adsorbed by the microorganisms in the contact zone, downstream of the feedpoint. The uptake of particulate substrate (less than 1.5 μm diameter) and colloidal substrate (0.03 μm < diameter < 1.5 μm) is rapid, while the uptake of soluble substrate proceeds more slowly (Bunch and Griffin, 1987). The microorganisms containing the adsorbed substrate are separated from the treated effluent in the secondary clarifier as in conventional activated-sludge WWTPs. Upon recycling to the aeration basin, stored substrate is metabolized in the stabilization zone upstream of the feedpoint, preparing the microorganisms for pollutant uptake in the contact zone.

Most municipal wastewaters contain a high percentage of either colloidal or particulate substrate and are suitable for treatment by either step-feed or contact stabilization, but industrial wastewaters must be evaluated on a case-by-case basis. However, it is anticipated that the degree of nitrification in WWTPs operated using step feed will be less than in equally sized, conventional activated-sludge WWTPs (operated at an equivalent MCRT) because of the reduced contact time between the soluble ammonia and the biomass (Georgousis *et al.*, 1993).

As illustrated in Figure 9.14, as the influent wastewater feedpoint is moved closer to the secondary clarifier, the solids loading to the clarifier is further reduced (Vaccari, 1986). However, as the point of influent wastewater addition is moved down the aeration train, the contact time between influent wastewater and biomass is reduced. At some point, there will be insufficient contact time, and substrate will pass through the system, resulting in poor effluent quality. While moderate effluent quality deterioration has been observed with step-feed operation (Hill, 1985), pilot- and full-scale experience treating domestic wastewaters has indicated that effluent suspended solids and biochemical oxygen demand (BOD) concentrations comparable to those achieved during conventional operation are maintained as long as the following criteria are satisfied (Thompson *et al.*, 1992a):

- Contact time between influent wastewater and biomass exceeds 1 hour, and
- Mixed liquor suspended solids concentration fed to the secondary clarifiers exceeds 1 g/L.

The actual effluent quality that can be achieved with step-feed operation is best determined experimentally at a given WWTP. The effect of step-feed operation on nitrification is likely to be more severe than that on organic removal, but this does not necessarily preclude the use of step feed for nitrifying WWTPs.

Step-feed operation alters the oxygen demand within the aeration basin. Frequently, the oxygen requirements downstream of the step-feed point increase because of increased bioactivity in the presence of high substrate concentrations. Tapered aeration systems may not be able to maintain a DO setpoint at the tail end of the aeration basin during step-feed operation. Effluent quality deterioration has been observed to be more pronounced when DO concentrations fall to less than 1 mg/L (Hill, 1985).

STEP FEED FOR STORM FLOW AND BULKING CONTROL. The primary application of step-feed operation is in the control of storm flow and sludge bulking (Torpey, 1948). Under high-flow conditions, mixed liquor solids are transferred from the aeration basins to the secondary clarifiers, increasing the depth of the sludge blanket until it potentially carries over the effluent weirs (sludge blanket washout). Susceptibility to washout is increased during periods of filamentous sludge bulking. To avoid washout, many WWTPs must bypass wastewater during high-flow conditions. Switching to step-feed operation reduces the solids loading to the secondary clarifiers, allowing a greater volumetric flow rate to be treated without sludge blanket washout.

Many WWTPs of varying sizes, including plants in Seattle and New York, routinely use step-feed operation. Generally, these WWTPs are operated continuously in a step-feed mode or converted to step feed before the wet weather season. They are not converted to step-feed operation for control of an immediate storm flow event.

Step-feed control of storm flow has been demonstrated at pilot scale (Thompson et al., 1989) and full scale (Thompson et al., 1992b). Typical results are presented in Figure 9.15. The plots display the response of the effluent suspended solids to storm flow for a plant operated with and without step-feed control. Demonstrated in Figure 9.15, bottom, without step-feed control, washout was observed within 2 hours of the storm flow entering the WWTP. Bypass was initiated to reduce the flows to normal levels to prevent further washout. In Figure 9.15, top, with step-feed control, the storm flow was treated throughout its duration with no bypass, and only a moderate increase in effluent suspended solids concentration was observed.

EXAMPLE CONTROL STRATEGY. A detailed investigation of step-feed control of storm flow was carried out at the Hamilton, Ontario, Woodward Avenue Water Pollution Control Plant (Thompson et al., 1992b). An operating strategy was developed that reduced bypass volume by an order of magnitude (Georgousis et al., 1992). Similar operating strategies can be developed at other WWTPs. The control strategy that was developed for the Woodward Avenue plant involves manual manipulation of motorized gates by operators, although the operators are provided with on-line data from flow, MLSS concentration, clarifier sludge blanket level, and effluent turbidity instruments. The control strategy involves the use of a storm flow capacity estimation program, STORMAX, developed for the Ontario Ministry of Environment and Energy.

The Woodward Avenue plant aeration basins consist of six partially baffled aeration cells in series. A dye test indicated that the aeration basins had plug-flow characteristics although there was significant backmixing. The plant was modified to allow primary effluent to enter the first, third, and fifth aeration cells or combinations of these. Motor-operated gates were installed to allow for rapid change to the influent wastewater feedpoint.

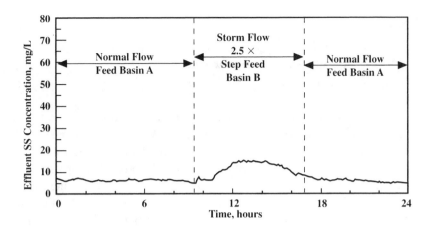

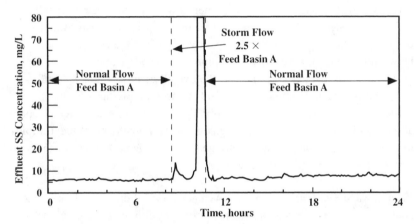

Figure 9.15   Effluent suspended solids concentration during storm flow with (top) and without (bottom) step-feed control (SS = suspended solids)

Before implementation of the control strategy, the secondary clarifier capacity was determined experimentally by increasing the surface overflow rate and monitoring performance (stress tests). Experiments were conducted over the historical range of MLSS and solids settling characteristics. From the data, a relationship was developed between the initial settling velocity of the mixed liquor and secondary settler capacity. The form of the relationship is based on the secondary settler design equation:

If: sludge $ISV > SF \times SOR$
Then: settlers are operated at less than capacity   (9.2)

If: sludge $ISV < SF \times SOR$
Then: settlers are operated at greater than capacity   (9.3)

Where

ISV = mixed liquor initial settling velocity;
SOR = clarifier surface overflow rate;
SF = safety factor.

Under dry weather conditions, conventional operation is maintained to ensure maximum nitrification. Once per week, WWTP operators conduct solids settling tests to obtain a relationship between initial settling velocity and MLSS concentration. The data are input into the predictive capacity model, STORMAX. For a given influent wastewater flow rate, the model calculates the resulting MLSS distribution for all potential methods of operation. It then applies the MLSS–settling-velocity relationship and calculates a capacity for each method of operation. Moderate step-feed action involves directing the influent wastewater flow to the third aeration cell, while severe step-feed operation involves directing influent wastewater to the fifth aeration cell. Finally, the model verifies that the contact time between the influent wastewater and biomass exceeds 1 hour.

There are two levels of control in the overall control strategy: feedforward and feedback. Feedforward control is based on predicted influent wastewater flow rates and predicted WWTP capacity for each mode of operation. Feedback control is based on sludge blanket depth and effluent turbidity. However, feedback control is not generally triggered because of the success of the feedforward control strategy.

Feedforward Control. During wet weather, WWTP operators have some control over the influent wastewater flow rate by filling storm flow retention tanks in the collection system. However, these tanks have limited capacity. To initiate the control strategy, operators estimate the influent wastewater pumping rate required to prevent overflow of the storm flow retention tanks. Secondly, they use STORMAX to determine the appropriate method of operation for this flow rate. If possible, operators convert to the required mode of operation at least 3 hours before increasing the influent wastewater pumping rate. Occasionally, they convert to a moderate form of step-feed operation based on a prediction of wet weather. Return activated sludge pumping rates are not adjusted.

Feedback Control. Effluent turbidity and the sludge blanket level in representative clarifiers are measured using on-line instrumentation. If excess turbidity is detected or if the sludge blanket approaches the effluent weirs, an alarm sounds. The operator confirms the alarm through visual observation and increases the degree of step-feed operation if it is required.

The control strategy allows for secondary bypass if it is required. Occasionally, under severe flow conditions or poor solids settling conditions, the entire storm flow cannot be treated even with the most extreme method of step-feed operation (all flow entering aeration cell 5). Secondary bypass is used to ensure that the contact time in the aeration basin between biomass and influent flow is maintained at greater than 1 hour at all times. The WWTP is converted back to lesser degrees of step-feed operation and finally to conventional operation as the influent wastewater flow subsides.

RETURN ACTIVATED SLUDGE CONTROL

Return activated sludge control is essential for proper performance of the secondary treatment processes: biological reaction and secondary clarification. The amount of RAS returned to the biological processes must be sufficient to maintain an adequate MLSS concentration to treat biologically the BOD and the nutrients in the influent stream. This requires that both the rate at which the influent BOD and nutrients enter the WWTP and the concentration of solids in the RAS be known.

Biochemical oxygen demand is the measure of the oxygen demand of the material in the influent flow stream. Biochemical oxygen deman measurement requires time-intensive laboratory analysis and is not suitable for continuous control methods because the results are obtained after the effect on the treatment process. The amount of BOD in the influent stream will vary throughout the day, somewhat proportional to diurnal flow variations.

A measurement for the microorganisms concentration in the mixed liquor is either total suspended solids (TSS) or volatile suspended solids (VSS). Both TSS and VSS require laboratory analysis. The concentrations of RAS TSS and VSS are affected by diurnal flow variations, performance and design of the aeration process, and the MLSS settling characteristics.

An integrated control strategy for RAS control should have the following objectives depending on the state of the system and desired result:

- Prevent gross process failure: no solids washout during high flow.
- Minimize suspended solids in the effluent.
- Prevent denitrification.
- Maximize thickening for solids processing.
- Prevent thickening failure caused by bulking solids or inadequate clarifier capacity.
- Optimize performance in terms of cost (energy, chemicals, labor, and other costs).

Several control strategies have been developed and are used successfully for RAS control to meet the above objectives. Operation of a WWTP may require choosing among different control strategies depending on influent conditions and desired results. Not all of the objectives can be met at once. These strategies require the WWTP operator to understand the dynamics and variables involved in RAS control (Keinath, 1985, and Keinath *et al.*, 1977). The strategies covered in this section are

- Fixed RAS flow control,
- Variable RAS flow control,
- Blanket level control,
- Return activated sludge flow control based on settled sludge method,
- Return activated sludge flow control based on solids flux theory, and
- Other advanced control strategies.

FIXED RETURN ACTIVATED SLUDGE FLOW CONTROL. A fixed RAS flow control strategy can be performed either manually by adjusting a valve position or pump speed to control the flow, or by automatic flow control. Figure 9.16 shows automatic fixed RAS flow control from a secondary clarifier to an aeration basin. The general strategy compares the RAS flow measurement to a setpoint and controls the RAS flow using feedback control. The feedback control signal is used to modulate either a variable-speed-driven pump or a control valve. Table 9.1 gives the minimum instrumentation for fixed RAS flow control.

This simple strategy works well and is used widely in WWTP operation. Fixed RAS flow control works well for minimizing suspended solids in the effluent and preventing denitrification if the RAS flow rate is high enough to be conservative and to keep blankets in the clarifiers low. Fixed RAS flow does not maximize thickening for subsequent solids processes. For varying influent flows, the fixed-flow control strategy produces variable MLSS in the aeration process, variable sludge blanket levels in the secondary clari-

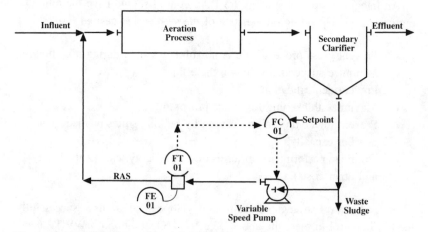

Figure 9.16   Return activated sludge control: fixed flow

Table 9.1   Minimum recommended instrumentation for return activated sludge control

Measurement	Comments
Fixed-flow control	
Flow rate	Return activated sludge flow measurement
Control element (valve or variable-speed pump)	Return activated sludge flow control
Variable-flow control	
Flow rate	Return activated sludge flow measurement
Flow rate	Influent waste flow measurement to aeration proces
Control element (valve or variable-speed pump)	Return activated sludge flow control

fier, and variable RAS concentrations, which may affect wasting strategies and downstream processes.

VARIABLE RETURN ACTIVATED SLUDGE FLOW CONTROL.
Variable RAS control is also known as *constant ratio* or *flow ratio* control. The variable RAS control strategy, shown in Figure 9.17, measures the RAS flow and the influent flow to the aeration process. The RAS flow is divided by the influent flow rate, and this ratio is compared to a setpoint. The output from the controller modulates a control valve or a variable-speed pump. The variable RAS flow control strategy assumes constant BOD or substrate concentration in the influent stream and constant solids concentration in the RAS. This strategy improves on fixed-flow RAS control in that it maintains a more constant RAS concentration and sludge blanket level in the secondary clarifier. Under high influent flows, typically caused during storm flows, variable RAS flow control may cause clarifier failure. Table 9.1 includes the minimum instrumentation for variable RAS flow control.

SLUDGE BLANKET LEVEL CONTROL.   The sludge blanket level control method is used to maintain a relatively constant sludge blanket level in secondary clarifiers. A minimum sludge blanket is maintained for efficient settling and to provide a relatively constant RAS concentration. The sludge blanket level is determined by periodic manual measurement or

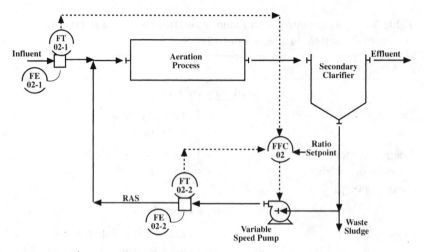

Figure 9.17 **Return activated sludge control: variable flow**

by using an automatic sludge blanket detector continuously. The RAS flow is adjusted based on the desired blanket level. Sludge blanket level control can be used for maximizing thickening processes.

High sludge blanket levels require an increase in RAS flow, and low sludge blanket levels require a decrease. In a WWTP running in a contact stabilization configuration, reducing the RAS flow rate will increase the MLSS concentration under reaeration and will reduce the MLSS concentration under aeration; the result will be a decrease in blanket height in the clarifier. Increasing the RAS flow rate will reduce the sludge blanket height in the short run, but increasing the RAS flow rate will ultimately increase the clarifier blanket height. Therefore, the use of sludge blanket level control should be coordinated with waste activated sludge control strategies because it will affect blanket levels.

RETURN ACTIVATED SLUDGE FLOW CONTROL BASED ON SETTLED SLUDGE METHOD. The settled sludge method is a measurement technique used to determine the fixed or variable RAS flow setpoint. A settleability test is used to estimate the volume of solids in the activated sludge flowing to the clarifier. A settleometer is used to determine the volume of settled sludge in 30 minutes. The following equation is used to determine the ratio of RAS to influent flow:

$$R/Q = SSV/(1\,000 - SSV) \qquad (9.4)$$

Where

R =   return activated sludge flow, m³/s (mgd);
Q =   secondary influent flow, m³/s (mgd); and
SSV =   settled sludge volume after 30 minutes, mL/L.

The *R/Q* ratio can be used as the setpoint in the variable-flow control strategy. For fixed-flow control, *R* can be calculated by multiplying *R/Q* by influent flow to the aeration process.

RETURN ACTIVATED SLUDGE FLOW CONTROL BASED ON SOLIDS FLUX THEORY. Return activated sludge flow control based on solids flux theory has been developed by Keinath and others (Keinath, 1985, and Keinath *et al.*, 1977). This method uses the state point concept of operating activated-sludge systems using the solid transport bulk flux and settling flux. Figure 9.18 shows a typical settling flux curve with the operating lines and state point. The two operating lines are the recycle flow rate operating line characterized by the slope of negative *u* and the clarifier overflow rate operating line characterized by slope *ORA*. The following equations are used to determine the slopes of the operating lines:

$$u = Q_R/A \qquad (9.5)$$

$$ORA = Q_i/A \qquad (9.6)$$

Where

Q_R = return activated sludge flow, m³/s (mgd);
A = clarifier surface area, m² (sq ft); and
Q_i = influent flow rate, m³/s (mgd).

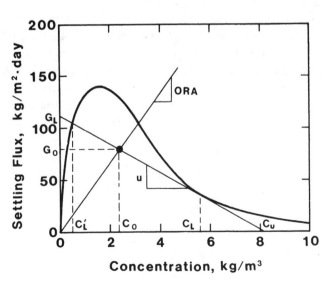

Figure 9.18 Typical settling flux curve with operating lines and state point (Keinath, 1985)

The state point x-coordinate C_o is the MLSS concentration, and the y-coordinate G_o is the solids flux $(C_oQ_i)/A$. The settling flux curve and operating lines can be used to set RAS flow rates as influent rates change. At the state point, and where the recycle flow rate operating line is tangent to the settling flux curve, the clarifier is underloaded. Changes in influent rate shift the state point vertically; changes in RAS flow rates change the slope of the recycle flow rate operating line. A clarifier that is overloaded for thickening has a state point below the settling flux curve and a recycle flow rate operating line that passes above the descending limb and below the maximum point of the settling flux curve. A clarifier that is overloaded for both thickening and clarification has a state point that is above the settling flux curve. If this condition persists too long, the clarifier will fail.

Keinath (1985) and Keinath and Stratton (1985) developed two strategies for coping with overloaded clarifiers during excessive influent flow rates: recycle rate control and step-feed control. Recycle rate control can be used successfully when the state point is below the maximum point of the settling flux curve. If the influent flow increases and the new operating state point is outside the envelope of the settling flux curve, the MLSS concentration is allowed to decrease, transferring solids to the clarifier. This is done by a two-step process. First, decreasing the RAS flow so the recycle rate operating line is tangent to the settling flux curve causes the state point to move down and to the left and to cross the settling flux line as the MLSS solids concentration decreases. After the state point crosses the settling flux curve, the RAS rate is increased so that the recycle rate operating line is again tangent to the settling flux curve.

For cases in which the state point is above the maximum point of the settling flux curve, recycle rate control will not work. Step-feed control can be used to prevent clarification failure if the process has been designed for changing aeration basin influent feed points. Otherwise, there is little that can be done except for changing hydraulic loading on the plant.

The settling flux curve for the WWTP must be determined for RAS flow control based on solids flux theory. This requires periodical determination of the settling flux curve by analytical and experimental methods. Daigger and Roper (1985) developed an empirical relationship using the sludge volume index (SVI) to determine the settling sludge curve. Furthermore, computer programs have been developed to assist the operator in analyzing operating states and specifying control actions (Keinath, 1985, and Keinath and Stratton, 1985).

OTHER ADVANCED CONTROL STRATEGIES. Other advanced control strategies include the specific oxygen uptake rate (SOUR) or expert systems that take into account the entire activated-sludge processes for control (Chandra *et al.*, 1987, and Lai and Berthouex, 1990). The SOUR method requires a mass balance for oxygen around the aeration tank and is instrument and maintenance intensive. The number of sensors and the difficulty of the measurements required for SOUR control are drawbacks to this method, and one of the other control strategies discussed above should be used as a fallback method (Chandra *et al.*, 1987).

SOLIDS WASTE CONTROL

Automatic control of the activated-sludge process waste flow rate is not yet commonly used. This is not because it is unfeasible, as current manual techniques could be automated. Part of the problem may be that operators and designers are skeptical of whether automatic control will result in any benefits, or even whether it will work at all. Clearly, there is a potential for important benefits, such as the ability to maintain the system closer to a desired state, with resulting reduction in permit violations and operating costs. Use of automatic control can allow a WWTP to operate closer to its design capacity by reducing variability in performance; this can reduce or eliminate the need for a plant upgrade. An examination of a variety of techniques that could be automated may help lead to a realization of the potential benefits.

This section will describe three parts of the problem of controlling wastage in the activated-sludge process: selection of a control objective, identification of a process, and determination of a control law. *Process identification* means determination of the process response to control actions and other inputs. The control law is a rule describing how to manipulate the variable to meet the control objective.

Waste rate manipulation has the overall goal of influencing the biology of the activated-sludge process. Although the biology can be influenced by other things such as DO, the wastage is generally considered to be a sort of "master variable" influencing the performance of the process. There are two lines of evidence for this. The first is based on the mathematical theories of microbial growth and substrate use. The second line of evidence is from both laboratory and field experience.

Both types of evidence depend on the ability to control total system biomass, MCRT, or F:M ratio. For a steady-state WWTP, these parameters are functionally related. All of them are determined by a balance between growth and wasting. Because microbes adjust their growth rates (within limits) to the amount of substrate available, only the waste rate remains as a "free" variable, available for operator manipulation.

The mathematical growth models show that the waste rate determines the MCRT (and therefore biomass and F:M) of a steady-state WWTP. The MCRT, in turn, is the inverse of the net growth rate. For example, a 10-day sludge age indicates a net growth rate of 0.1 per day; that is, the biomass reproduces at a rate of 10% of its mass per day. Now, one of the

---

basic assumptions of microbial growth theory is that microbial growth rate depends on the concentration of substrate (BOD) available. By setting the growth rate, the operator in effect sets the effluent soluble BOD. However, most of the BOD in the effluent is often particulate because of the escape of solids particles from the clarifier.

In the second type of evidence for waste rate control, it has been found that particulate BOD is also related to MCRT. This is because of the fundamental effect that biomass, MCRT, and F:M have on the biology of the system. By setting the net growth rate, some microbes compete more successfully than others, whereas slower-growing organisms are washed out altogether. The resulting change in the mixed liquor concentration and the SOUR affects the DO in the aeration basin and changes the DO distribution in the interior of the solids flocs. These and other effects influence the size and density of the resulting flocs, thereby affecting effluent suspended solids. In the majority of WWTPs, the MCRT required to produce a well-settling sludge is substantially greater than the MCRT required to remove essentially all of the soluble BOD.

There is now an improved understanding of the effect of parameters such as DO and loading rate on specific organisms, such as filamentous growth. Nevertheless, the relationship between the physical environment and the microbial population is mostly an individualized problem for each WWTP. It is the job of the conscientious and observant operator to determine the conditions that work best at a particular WWTP.

CONTROL OBJECTIVES. In selecting and using a control strategy for the activated-sludge process, the first task is to carefully specify the *control objective*. An objective is some parameter, or a function of one or more parameters, that is to be either minimized or maximized or may act as a constraint. For example, it may be desired to minimize solids disposal costs (where beneficial use is not an option) by operating at a high MCRT (resulting in high MLSS). However, the MLSS will be constrained by the ability of the aeration system to provide enough oxygen and by the ability of the clarifier to remove the solids from the effluent and return them to the aeration tank. Such constraints are often fixed by the design of the WWTP.

When this question is examined more carefully, it is seen that numerous objectives can be stated and that they may even contradict one another. Some parameters that may be included in control objectives include effluent suspended solids, effluent BOD (soluble and total), effluent ammonia, solids production, aeration tank DO, clarifier solids loading rate, clarifier blanket level, and growth of filamentous bacteria or other nuisance organisms.

The effluent suspended solids and BOD are, of course, the main performance variables, as well as ammonia in WWTPs concerned with nitrification. Often, the objective is to maintain these variables below the levels specified in the WWTP's discharge permit, at lowest cost, and with some margin of error. However, the operator may also have a responsibility to the public to try to reduce the level of pollution in the effluent, even if standards are being met.

Solids production minimization relates to the issue of cost of disposal when beneficial use is not an option. Solids generation is easily minimized by operating at the longest feasible MCRT—in other words, by using the lowest possible waste rate. This may result in problems with one or more other objectives, such as effluent quality, clarifier solids loading, energy use, or aeration tank DO. In this case, these other objectives would be formulated as constraints. Note that the optimum MCRT in terms of cost might not be the longest feasible sludge age because the cost of aeration may go up with the MCRT. It is possible for energy costs to increase faster than disposal costs. Obviously, this will be sensitive to changes in disposal costs and methods. Also, the operator should consider the effect on the entire WWTP; increases in solids production might be balanced by increased destruction and gas formation in anaerobic digesters, if these are present. In all cases, the possible beneficial use of the generated biological solids (biosolids) should be considered the more environmentally sound design option (versus disposal) unless cost or other factors mitigate against it.

Day-to-day variations in aeration tank DO and clarifier solids loading rates should be controlled using air flow and return sludge flow (or step-feed configuration), respectively, as described elsewhere in this manual. However, too high an MCRT could reduce the "breathing room" for these control actions. As mentioned above, there is a designed-in MCRT limitation to the process. If optimum MCRT in terms of settleability violates a DO or clarifier loading constraint, then the WWTP may be a candidate for an upgrade, such as by installation of additional blower capacity, conversion to pure oxygen, or construction of additional clarifiers.

Clarifier sludge blanket level should typically be controlled by the return sludge flow rate. Occasionally, waste flow rate may be used to maintain a sludge blanket in the clarifier. This would only be feasible if the influent flow rate is fairly constant, such as at small WWTPs with flow equalization or in industrial facilities, or if control is performed either automatically or manually with frequent (hourly) monitoring and control. Although this allows operation at a high MCRT or with a smaller WWTP, use of this strategy should be accompanied by careful attention to effluent quality. The clarifier should be monitored for rising solids or septic conditions.

Return flow rate and step-feed pattern should not be used to control MCRT. Wastage should be controlled only by manipulating the waste flow rate. Although return rate and step-feed pattern have an effect on wastage, their real purpose is to control solids distribution. Separating these effects from wasting is termed *decoupling*.

By manipulating growth rates (indirectly) and environmental conditions in the aeration basin, waste rate control can exert an influence on the microbial ecology of the activated-sludge system. In this way, specific organisms or groups of organisms can be selected for or against. In the case of filamentous bulking, it is possible to identify the type of organism causing the bulking and use this information to determine the type of process control change to bring the bulking under control. This kind of strategy is beyond the scope of this text.

All of the objectives mentioned above are of direct interest to the operator; that is, if one of them is out of control, that in itself would be a problem. Unfortunately, the first two, effluent suspended solids and total BOD, are not directly influenced by the waste rate. Effluent soluble BOD is influenced by the MCRT and, thus, the solids waste rate. However, it does not present a measure of the overall system performance; thus, its value is limited. As a result, the operator seeks to control surrogate parameters that have an indirect effect on suspended solids and BOD. These include biomass, MCRT, F:M, SOUR, and SVI. Assuming steady state (no change in any flows or concentrations in the system), all of these except for SVI are theoretically equivalent to each other as control objectives. For dynamically changing WWTPs, each can vary somewhat independently.

The behavior of the activated-sludge microorganisms is affected by all of these parameters. The activated-sludge process is a starved culture of microorganisms. The low substrate concentration, which of course is desired from an effluent standpoint, has two results. First, it greatly decreases the microbial growth rate from the maximum rate when food is not limited. Second, under the competitive conditions of starved growth, the bacteria form flocs to improve their ability to capture and digest food. This has the benefit from an operational point of view of making it possible to remove the bacteria from the effluent by gravitational settling. However, once the system is maintained at an MCRT sufficiently above a minimum value (approximately 1 to 2 days), flocculation and substrate concentration are rarely a problem. The problem that remains is to select for organisms that will produce a well-clarified effluent and easily thickened (nonbulking) solids.

Whether the problem is settling or BOD removal, the operator needs to determine the relationship between the surrogate variables and effluent performance. This is one of the key problem-solving areas in the activated-sludge process. This will be discussed in the section below on Process Monitoring and Correlation, after a discussion of the individual control parameters.

The biomass (or M) in the system is the simplest and most readily determined parameter for wastage control. Its effect on the properties of the solids in the system occurs through its relationship with growth rates and with the growth environment in terms of MLSS and DO. There is a point of controversy among operators concerning measurement of biomass and whether it is necessary to include solids stored in the clarifier. If it is not considered, then

$$M = X_a \times V \qquad (9.7)$$

Where

X_a = the average of the suspended solids in the aeration tank, mg/L; and

V = the total aeration tank volume, m^3.

The justification for using this equation is that, because the solids in the clarifier are not being aerated, growth slows and the bacteria approach an inactive state. This approach is acceptable if the solids in the clarifier represent a relatively small fraction of the total (typically true if a sludge blanket is not being maintained) and X_a is measured at a time of day when clarifier sludge is at a minimum (when influent flow is low, as in the early morning).

A disadvantage with this approach is that, if solids shift to or from the clarifier because of flow or settleability changes, M will seem to change even if it has not. This is particularly important in the calculation of F:M and MCRT, as discussed below. It can also be argued that microbial activity varies throughout the process, from feed point to return sludge line, which can be seen in the wide changes in SOUR at different locations in the aeration basin. Although oxygen uptake slows at low DO, other metabolic activities might not (for example, glycolysis).

A better approach to eliminate possible inconsistencies is to measure the average concentration of solids in the clarifier (X_c) using tube samplers with a check valve in the bottom, which are commercially available. Then, biomass is computed as follows:

$$M = X_a \times V + X_c \times V_c \qquad (9.8)$$

Where

V_c = total volume of the clarifiers, m^3.

The MCRT, or sludge age, is one of the most popular parameters for control. It is typically computed assuming that the WWTP is in steady state, which is rarely the case for real plants. A more realistic calculation, the dynamic sludge age (DSA), is described below. The MCRT represents the average time a particle of solids spends in the system between the time it is produced and the time it is removed by wasting or loss in the effluent. Thus, the MCRT, coupled with the growth rates of organisms, acts to select the microbial ecology of the system. The MCRT in a dynamic system represents a balance between growth and wasting over a period of time. There-fore, it may be considered to be a parameter that describes in some sense the recent "history" of the process in terms of growth and wasting.

The F:M ratio, conversely, represents what is currently being done to the process. It is the number of grams of BOD being provided for each kilo-gram (pounds per pound) of microorganisms per day. Again, as long as the mass in the clarifier remains a small and unchanging fraction of the total, it can be ignored. Otherwise, it is essential to include clarifier mass.

Similar arguments for including clarifier mass apply here as for the MCRT calculation. Although solids in the clarifier are not currently being loaded with BOD, neither are solids that have left the first pass of the aera-tor (in a conventional plant). The solids recirculate through the entire sys-tem several times per day, and all solids are exposed to influent each time. Therefore, inclusion of clarifier mass gives the best measurement of the daily amount of substrate provided for each kilogram (pound) of biomass.

PROCESS MONITORING AND CORRELATION. An important part of the process control task is to observe the process and to determine the relationships among various qualitative and measured parameters. Qualita-tive parameters include observations of color, odor, appearance of aeration tank surface, and microscopic examination. These factors can be assigned numerical scales for purposes of graphical and statistical checking described below. For example, amount of filamentous growth can be rated from 1 to 5. Measured parameters include (in addition to the control objec-tives and surrogate parameters listed above) influent BOD and suspended solids concentrations and loading rates, and SVI.

Mass Balance Parameters. The most important tool in the quantitative monitoring of the activated-sludge process is the mass balance. The useful-ness of the MCRT, F:M, and biomass measurements depends on the accu-racy of the mass balance. Flow-proportioned sampling of all influent and effluent flows, together with periodic (daily) sampling of the entire system volume, ensures the necessary accuracy. Including the clarifier in the daily sampling reduces unaccountable changes in biomass. In the following, all flows are assumed to be averages over a specified time period (typically 1 day), and concentrations are flow-proportioned averages.

The F:M ratio is computed from the BOD loading rate:

$$\text{BOD loading rate} = (Q \times S_0) \tag{9.9}$$

Where

Q = influent flow rate, m³/d; and
S_0 = organic concentration entering the activated-sludge process, mg/L.

The influent organic concentration can be the BOD, although this would not be useful for automatic control. Any surrogate that can be measured on line, such as total organic carbon, can substitute. Then

$$F:M = \frac{S_0 \times Q}{M} \qquad (9.10)$$

A similar parameter to the F:M is the substrate utilization rate (U):

$$U = \frac{Q \times (S_0 - S)}{M} \qquad (9.11)$$

Where

$$S \ = \ \text{effluent BOD, mg/L.}$$

The substrate utilization rate more accurately represents the food actually used by the microorganisms. However, because S is typically much less than S_0, F:M and U are almost identical.

It is also be useful for WWTPs to monitor their solids production rate (P). First, it is necessary to compute the total rate of mass wasting from the system (W):

$$W = (Q - Q_w) \times X_e + Q_w \times X_w \qquad (9.12)$$

Where

$$\begin{aligned}
Q_w &= \text{average waste sludge flow rate, m}^3; \\
X_w &= \text{waste sludge volatile suspended solids, mg/L; and} \\
X_e &= \text{effluent volatile suspended solids, mg/L.}
\end{aligned}$$

This equation applies whether wasting is from mixed liquor or from return sludge. Next, the accumulation rate of solids in the system (K) is computed over a 24-hour period using the mass in the system at the beginning of the period (M_0) and the mass at the end of the period (M):

$$K = \frac{(M - M_0)}{t} \qquad (9.13)$$

Where

$$t \ = \ \text{length of the time period (typically 1 day, although sometimes spanning a weekend).}$$

Finally, the solids production rate (P) is the sum of the solids waste rate and solids accumulation rate:

$$P = W + K \qquad (9.14)$$

Of interest to the operator is not simply the production rate, but the specific solids production ratio, P/M. The specific solids production ratio bears the same theoretical relationship to U for dynamic WWTPs as $1/\theta_c$ (Hiraoka and Tsumura, 1989, and Vaccari and Christodoulatos, 1990), where θ_c is the MCRT:

$$1/\theta_c = Y \times U - k_d \qquad \text{(steady state)} \qquad (9.15)$$

$$P/M = Y \times U - k_d \qquad \text{(unsteady state)} \qquad (9.16)$$

Where

Y = microorganism yield, mg X/mg S; and
k_d = decay coefficient, $1/d$.

Thus, at steady state, $1/\theta_c = P/M$, because $P = W$ in this case. The unsteady-state relation can be used to estimate Y and k_d for a particular WWTP. A simpler calculation is to compute the observed yield (Y_{obs}):

$$Y_{obs} = \frac{P/M}{U} \qquad (9.17)$$

Thus Y_{obs} can be computed on a daily basis. The meaning of Y_{obs} is essentially the amount of solids produced per quantity of BOD removed. Changes in this parameter may indicate changes in microbial activity or in influent quality.

Calculation of Mean Cell Residence Time. The MCRT is typically defined in terms of the calculation for steady-state processes, also called the traditional sludge age (TSA) (Vaccari et al., 1985):

$$\text{TSA} = \frac{M}{W} \qquad (9.18)$$

However, because real WWTPs are not at steady state, the TSA calculation causes problems. It is sensitive to day-to-day variations in W. The TSA may even predict that the sludge "ages" more than 1 day in a 24-hour period. For example, if W is cut in half, the TSA doubles. This does not realistically describe any changes in the biomass. As another example, an increase in BOD loading, other things kept equal, does not cause any

Automated Process Control Strategies

change in TSA because wasting increases in proportion to mass. In reality, the MCRT initially decreases because of an increased amount of new solids production, then gradually approaches the steady-state value (Vaccari *et al.*, 1985).

Dynamic sludge age, an improved calculation, accurately computes the true average MCRT for unsteady-state systems. The DSA calculation is more complicated than that of the TSA, but with the availability of personal computers this is not a serious drawback. A computer program for its computation is available (Vaccari *et al.*, 1988).

Calculation of DSA at the end of a time period (*t*) requires the initial value (DSA$_0$), which is the MCRT at the time M_0 was measured. When first starting the calculation, the previous day's TSA can be used; any error decreases with time. There are four special cases for computing DSA:

Case 1) $W = 0$; no solids were wasted or lost over period *t*:

$$\text{DSA} = (\text{DSA}_0 + t) \times \left(\frac{M_0}{M}\right) + \left(1 - \frac{M_0}{M}\right) \times \frac{t}{2} \qquad (9.19)$$

Case 2) $K = 0$; there was no net solids production ($W = P$):

$$\text{DSA} = \frac{M_0}{W} + \left(\text{DSA}_0 - \frac{M_0}{W}\right) \times \exp\left(\frac{-W \times t}{M_0}\right) \qquad (9.20)$$

Case 3) $W = 2P$; exactly twice as much solids are wasted as were produced:

$$\text{DSA} = \text{DSA}_0 \times \frac{M}{M_0} - \left(t + \frac{M_0}{K}\right) \times \ln\left(\frac{M_0}{M}\right) \qquad (9.21)$$

Case 4)   None of the above conditions hold, then:

$$\text{DSA} = \left(\text{DSA}_0 - \frac{M_0}{P + K}\right) \times \left(\frac{M}{M_0}\right)^{(-P/K)} + \frac{M}{P + K} \qquad (9.22)$$

Most of the time, case 4 will hold; however, the conditions must be checked first to determine if one of the other equations should be used instead.

The TSA parameter still has meaning as the ultimate MCRT toward which the system is tending, or as just a ratio between mass and wastage.

In summary, the operator should consider monitoring the following measured parameters and try to find relationships between them and process

---

Activated-Sludge Treatment   **129**

performance: M, F:M (or U), P/M, Y_{obs}, TSA, and DSA, as well as SOUR and DO.

Graphical Analysis. The most powerful tool for finding relationships among process variables is the operator's eye, aided by graphical display of data. This should be done even if automatic control is implemented because it can flag changes in process behavior that may require retuning or modification of control laws. Graphical analysis is easily done using a personal computer. Specialized software is available to assist in the job, but the commonly available spreadsheet programs are well suited to the task.

The operator should periodically examine the data from numerous points of view. Graphic analyses in the form of *X-Y* scatter-plots should be generated for

- All variables versus time (called *trend plots* or *time-series plots*); and
- Performance variables (for example, effluent BOD or SVI) versus all of the possible control variables (such as MCRT and biomass) and versus input variables (influent BOD).

Each comparison should be made over several time scales. That is, plots of each relationship should be made for periods covering the last several weeks, the last several months, and the last several years. The shorter time intervals should be plotted more frequently.

Patterns in the data should be noted. Their possible effects on performance should be considered. Occasionally, the data may suggest changes in the process control strategy. Notes and plots from periods of process upset may be useful during future upsets.

Statistical Analysis. As powerful as it is, graphical analysis has several limitations. It does not lend itself to making predictions or control laws; it may miss trends that have two or more simultaneous causes; and it may be difficult to detect by eye whether some slight effect is significant or just a chance occurrence. As with graphical analysis, statistics can be used to monitor the process even if automatic control is being used, and in fact, statistics may be incorporated into the control software.

Several statistical techniques are available that can overcome the drawbacks of graphical analysis. One is the use of control charts for tracking trends. The other method is the use of regression to determine cause and effect relationships, to make predictions, and to formulate control laws. These methods will be described only briefly here; references are given for further details.

Shewhart control charts are a way of combining statistics with the graphical analysis described above. In its basic form, the Shewhart control chart is the same as the time-series plot, but with upper and lower *control limits* drawn on the plot. The control limits are computed as the average value of the parameter being plotted, plus and minus two standard deviations of the parameter (Wadsworth, 1990). Note that often, as in the case for effluent BOD and suspended solids, it is necessary to compute the control limits using a logarithmic scale. If the value of the parameter on a particular day falls outside the control limit, that is a signal to the operator to search for potential problems to bring the process back under control. This method is intended to provide early warning for such excursions so that the operator can head off further degradation in performance.

There is a technical problem with using Shewhart charts for monitoring daily performance data: the basic Shewhart chart method assumes that the data vary in a random way from day to day. However, WWTP performance data on a particular date typically show strong dependence on the previous day's value. This dependence is called *autocorrelation*. Ignoring autocorrelations may result in missing significant excursions or flagging an excursion that is not significant. A method for applying Shewhart charts to autocorrelated data from WWTPs is given by Berthouex (1989).

Regression analysis is a technique for identifying a mathematical equation that gives the relationship between a dependent variable (such as a performance variable) and one or more independent variables. The independent variables are those considered to be causes of WWTP performance, such as MCRT. Thus, this is a technique that can identify multiple causes, and the resulting equation can be used for predictions and for process control.

A beginning approach to using regression with WWTP data was given by Perkins (1990). This reference shows how to find multilinear relationships. However, autocorrelations and nonlinearity can cause problems. Novotny *et al.* (1991) and Christodoulatos and Vaccari (1992) provide methods for handling autocorrelations. They also show how to handle nonlinearity in the data (Vaccari and Christodoulatos, 1992). A good reference that covers the basic statistics involved is Draper and Smith (1981). Although somewhat complex, all of these statistical methods can be applied using the same spreadsheet software as can be used for the graphical analysis.

CONTROL STRATEGIES. Having implemented process monitoring and established control objectives related to solids wasting, the remaining problem is to implement a control strategy for actually setting the waste flow rate. The objectives considered here will be control of MCRT, mass, or F:M.

Decoupling Interactions. In controlling MCRT, mass, or F:M, it is the mass waste rate (kilograms [pounds] per day) that is fundamental. The operator physically controls the waste flow rate. Therefore, the waste flow rate should be compensated for anything that affects the waste solids concentration, X_w. Factors that affect X_w include influent and return flow rate and step-feed configuration. The effect of step feed is discussed elsewhere in this manual.

The effect on waste solids concentration of influent flow and return flow are found from an approximate mass balance around the clarifier, neglecting solids loss with the effluent. The mass flow entering with the feed is then equal to that in the return line:

$$X_{ml} \times (Q + Q_r) = X_r \times Q_r \qquad (9.23)$$

Where

r =   recycle ratio;
r = Q_r/Q; and
X_{ml} =   MLSS, mg/L.

This is solved for X_r (in the case of return solids wasting):

$$X_r = X_{ml} \times \left(\frac{Q + Q_r}{Q_r}\right) = X_{ml} \times \left(\frac{1 + r}{r}\right) \qquad (9.24)$$

In a conventional or complete-mix activated-sludge process, X_{ml} is approximately independent of influent or return flow at a given amount of mass in the system. However, if wasting is done from the return line, the above equation shows that adjustment must be made to the waste flow whenever the recycle ratio changes, whether because of operator action on Q_r or uncontrolled changes in Q. This does not mean that Q_w needs to be adjusted hourly for influent flow changes; it is sufficient to compensate only for changes in daily average values.

Some operators use changes in return rate to manipulate MCRT. For example, reducing Q_r increases X_r. If return sludge is wasted at the same flow from the return line, more solids will be wasted per day and the MCRT will decrease. This is an inappropriate use of return flow rate; the proper purpose of Q_r is to control solids in the clarifier. Altering it for MCRT control can cause other, independent effects such as formation of a sludge blanket or even clarifier overloading.

Constant Waste Flow Rate. The simplest strategy is to keep the waste flow rate constant, only occasionally making adjustments when control or performance variables wander outside a specified range (or outside the Shewhart control limits). This is appropriate for WWTPs that do not experience extreme variations in influent quantity or quality and are not operating close to their design limits.

This method is also appropriate if the control objective is to set mass or F:M. These parameters do not have simple relationships to waste rate as MCRT does (below). The actual mass or F:M obtained with a given waste rate depends on Y_{obs}, which is variable and may change with time. If a value of Y_{obs} has been computed as described above, the value of waste flow required for a given mass can be estimated (neglecting solids in the clarifier and effluent suspended solids and BOD):

$$Q_w = Y_{obs} \times \frac{Q_{wL}}{Q + Q_r} \times \frac{V \times S_0}{M} \qquad (9.25)$$

Where

Q_r = return sludge flow rate, m³; and

Q_{wL} = flow rate in the line or basin from which solids are wasted, m³.

That is, $Q_{wL} = Q + Q_r$ if mixed liquor is wasted, and $Q_{mL} = Q_r$ if return sludge is wasted. S_0 is the daily average influent BOD. This equation can be used to adjust an out-of-bounds M or F:M.

Constant Proportion. A relationship can be derived among wasting, mass, MCRT, and other measurable parameters, based on a steady-state mass balance. When solved for waste flow, it yields a control law (Keinath and Cashion, 1980):

$$Q_w = \frac{M/\theta_c - Q \times X_e}{X_w - X_e} \qquad (9.26)$$

This equation gives the flow rate that will waste a constant proportion of the mass in the system, ultimately resulting in a desired MCRT. It is used to adjust the waste flow rate on a daily basis. This equation should only be used with average daily values.

Hydraulic Wasting. If mixed liquor is being wasted in a conventional or complete-mix activated-sludge process, then $X_w = X_{ml}$, and if clarifier mass can be neglected, $M = V \times X_{ml}$. If, in addition, solids loss in the effluent can be neglected, the equation for proportional wasting reduces to

$$Q_w = \frac{V}{\theta_c} \qquad (9.27)$$

This simple relationship shows that a constant waste flow rate automatically yields a fixed (steady-state) MCRT. This is called *hydraulic wasting* (Garrett, 1958).

Hydraulic wasting can be extended to WWTPs that waste return sludge and to step-feed and contact stabilization plants. For example, in the case of return sludge wasting,

$$Q_w = \frac{V}{\theta_c} \times \frac{r}{1 + r} \qquad (9.28)$$

Thus, in WWTPs that waste return sludge, the waste flow rate also sets the MCRT, as long as the recycle ratio does not change.

Feedback Control. The control laws above use the desired, or setpoint, values of MCRT to compute the required waste flow. Feedback control uses the setpoint and the actual currently measured value to compute a waste rate to reduce the difference, or error, between the two. For example, a simple proportional feedback controller for mass would set the waste flow rate according to the following equation:

$$Q_w = Q_w' + K_c \times E \qquad (9.29)$$

where $E = M - M'$ is the error and M' is the mass setpoint. Q_w' is the waste flow used when the error is zero, and K_c is called the *controller gain*. A more general form of feedback control is the proportional-integral-derivative (PID) controller. This type of controller is discussed in Chapter 3.

Feedback control has not been applied to manipulation of activated-sludge waste rate to date (Vaccari *et al.*, 1988). Research results have shown that feedback control is capable of more closely controlling MCRT or mass than are conventional control methods (Vaccari and Christodou-latos, 1989). Other practical problems have not yet been addressed. For example, controlling waste mass flow rate would increase the variability in waste flow rate, affecting solids-handling processes. However, the practicality of implementing such a controller will depend more on solving the problem of identifying the relationship between process performance and such surrogate parameters as MCRT.

Feedforward (Ratio) Control. Feedforward control attempts to manipulate the system before changes caused by inputs can lead to an error. In the case of MCRT (as computed by DSA) a feedforward controller cannot be designed because the MCRT responds more quickly to changes in inputs than it does to changes in waste flow rate.

A simpler version of feedforward control is ratio control. It is similar to proportional feedback control except that the waste rate would be set proportional to some input instead of some result. For example, the waste flow rate could be set proportional to influent organic loading rate:

$$Q_w = K_c \times (Q \times S_0) \qquad (9.30)$$

Adjustments could also be made for changes in X_w caused by, for example, return ratio. A surrogate for influent BOD that could be measured rapidly is needed for this strategy.

Comparison of Control Strategies. The different strategies described above were studied by means of computer simulation that included the effects of diurnal flow variations and daily changes in flow and BOD obtained from a regional WWTP (Vaccari et al., 1988). The study showed how well each strategy would work and what the effect of controlling one parameter would be on the other parameters. Table 9.2 shows the results. The mean and standard error are shown for DSA, mass, and F:M. All of the controllers were adjusted for a nominal MCRT of 5.5 days. In the case of PID control of mass, the setpoint was 106 800 lb (48 400 kg), which corresponded to a 5.5-day MCRT. The proportionality on the ratio controller was 40 gal wasted per pound (333 L/kg) of BOD loading. The PID and ratio controllers acted on an hourly basis; the other controllers were limited to daily manipulation of waste rate.

Table 9.2 Means and standard errors resulting from differing control strategies (F:M = food-to-microorganism ratio; lb × 0.453 6 = kg)

Control strategy	Dynamic sludge age (days)		Mass (1 000 lb)		F:M (per day)	
	Mean	Standard error	Mean	Standard error	Mean	Standard error
Constant flow						
Mixed liquor	5.44	6.9%	107	12.1%	0.221	24%
Return sludge	5.65	7.8%	102.8	13.5%	0.231	24%
Constant proportion						
Mixed liquor	5.32	6.8%	95.2	13.2%	0.248	23%
Return sludge	5.51	6.7%	100.4	13.0%	0.236	23%
Feedback						
Dynamic sludge age	5.30	5.5%	107.5	21.2%	0.226	27%
Mass	5.49	12.3%	106.8	0.3%	0.221	25%
Feedforward						
Ratio	5.38	10.7%	104.7	4.7%	0.226	24%

The constant-flow and constant-proportion strategies represent methods of control that can be implemented manually. The feedback and feedforward controllers require automated equipment including sensors, computers, and actuators.

The standard deviation shows how close the controller kept to the mean. Table 9.2 shows, for example, that PID control of DSA produced the smallest variation in DSA (standard error = 5.5%) but the greatest variation in mass (standard error = 21.2%). Conversely, PID control of mass reduced variation in mass to a negligible amount but caused the widest variation in DSA. This illustrates how control objectives could conflict and highlights the need to establish how an actual WWTP responds to the various parameters. Note that ratio control obtained the best control of mass, other than that obtained by PID control of mass.

Among the constant-flow and constant-proportion methods, constant-flow control of return sludge wasting was slightly worse than the others. This method resulted in slightly higher variability in both DSA and mass, as compared to the other manual methods. Constant-proportion wasting performed slightly better for the case of return sludge wasting than did constant flow. Constant-proportion wasting of mixed liquor was also tried with hourly adjustments of waste flow rate instead of daily adjustments. This resulted in parameter variability similar to the daily constant-proportion strategies.

The variation in F:M was large for all control strategies studied. Dynamic sludge age feedback produced the largest F:M variability by a small margin. Apparently, most of the variation was caused by the relatively rapid changes in BOD loading versus the slow changes in mass. Thus, F:M ratio can only be controlled in the long term.

Downstream Effects. Some WWTPs may be limited by their solids-handling capacities downstream of the activated-sludge process. In this case, it may be necessary to consider the effect of the wasting strategy on downstream processes, in particular on a thickener. In general, any such process could operate closest to its capacity if the variability in loading were minimized. In this case, the best strategy is obviously hydraulic wasting from mixed liquor. A potential disadvantage of mixed liquor wasting is that Q_w is higher than it would be for return wasting because of the lower concentration wasted.

If return sludge is wasted, then the volumetric loading is smaller, but the solids loading rate is more variable. If this causes problems with solids overloading in the thickener, then it may be necessary to reduce Q_w during the day, when X_r is high, and increase it at night. Operating a WWTP in a step-feed mode decreases the variability in return sludge concentration.

CONCLUSION. As stated at the beginning of this section, automatic control of waste flow is full of potential that has not yet been realized. An impediment to implementation is that the relationship between process control parameters and process performance has not been completely reduced to theory. The relationship is known in the form of the knowledge of expert operators. It may be possible to capture this knowledge for automatic control by use of advanced systems, perhaps in combination with statistical techniques. In any case, it is certain that the experienced and knowledgeable operator will never be out of the loop but, rather, will always be an integral part of the process.

REFERENCES

Astrom, K.J., and Wittenmark, B. (1984) *Computer Controlled Systems— Theory and Design.* Prentice-Hall, Englewood Cliffs, N.J.

Berthouex, P.M., (1989) Constructing Control Charts for Wastewater Treatment Plant Operation. *J. Water Pollut. Control Fed.*, **61**, 1534.

Buhr, H.O., *et al.* (1984) Making Full Use of Step Feed Capability. *J. Water Pollut. Control Fed.*, **56**, 325.

Bunch, B., and Griffin, D.M. (1987) Rapid Removal of Colloidal Substrate From Domestic Wastewaters. *J. Water Pollut. Control Fed.*, **59**, 957.

Chandra, S., *et al.* (1987) Evaluation of Oxygen Uptake Rate as an Activated Sludge Process Control Parameter. *J. Water Pollut. Control Fed.*, **59**, 1009.

Christodoulatos, C., and Vaccari, D.A. (1992) Correlations of Performance for Activated Sludge Using Multiple Regression with Autocorrelation. *Water Res.* (G.B.).

Considine, D.M. (1985) *Process Instruments and Controls Handbook.* 3rd Ed., McGraw-Hill, Inc., New York, N.Y.

Coughanowr, D.R., and Koppel, L.B. (1965) *Process Systems Analysis and Control.* McGraw-Hill, Inc., New York, N.Y.

Daigger, G.T., and Roper, R.E., Jr. (1985) The Relationship Between SVI and Activated Sludge Settling Characteristics. *J. Water Pollut. Control Fed.*, **57**, 859.

Draper, N.R., and Smith, H. (1981) *Applied Regression Analysis.* 2nd Ed., Wiley, New York, N.Y.

Flanagan, M.J., and Bracken, B.D. (1977) *Design Procedures for Dissolved Oxygen Control of Activated Sludge Processes.* EPA-600/2-77-032, U.S. EPA, Cincinnati, Ohio.

Garrett, M.T. (1958) Hydraulic Control of Activated Sludge Growth Rate. *J. Water Pollut. Control Fed.*, **47**, 1055.

Georgousis, Z., *et al.* (1992) Step Feed Control of Storm Flow: Capacity Estimation. Paper presented at the 65th Annu. Conf. Water Pollut. Control Fed., New Orleans, La.

Georgousis, Z., *et al.* (1993) The Feasibility of Implementing Step Feed Control of Storm Flow at Selected Water Pollution Control Plants: Phase II. Report prepared for the Ontario Ministry of Environment and Energy, Water Resources Branch, Toronto, Ont., Canada.

Hermanowicz, S.W. (1987) Dynamic Changes in Populations of the Activated Sludge Community: Effects of Dissolved Oxygen Variations. *Water Sci. Technol.,* **19,** 889.

Hill, R. (1985) Dynamics and Control of Solids Liquid Separation in the Activated Sludge Process. Ph.D. dissertation, Rice University, Houston, Tex.

Hiraoka, M., and Tsumura, K. (1989) System Identification and Control of the Activated Sludge Process by Use of a Statistical Model. *Water Sci. Technol.,* **21,** 1161.

Keinath, T.M. (1985) Operational Dynamics and Control of Secondary Clarifiers. *J. Water Pollut. Control Fed.,* **57,** 770.

Keinath, T.K., and Cashion, B.S. (1980) *Control Strategies for the Activated Sludge Process.* EPA-600/2-80-131, U.S. EPA, Washington, D.C.

Keinath, T.M., and Stratton, S.C. (1985) Computer Assist For Activated Sludge Process. *Water Eng. Manage.,* 30.

Keinath, T.M., *et al.* (1977) Activated Sludge—Unified System Design and Operation. *J. Environ. Eng. Div.,* **103,** 829.

Kulin, G., *et al.* (1983) Evaluation of a Dissolved Oxygen Field Test Protocol. *J. Water Pollut. Control Fed.,* **55,** 178.

Lai, W., and Berthouex, P.M. (1990) Testing Expert Systems For Activated Sludge Process Control. *J. Environ. Eng.,* **116,** 890.

Liptak, B.G., and Venczel, K. (1982) *Instrument Engineers Handbook: Process Measurements.* Chilton Book Company, Radnor, Pa.

Lutman, C.G., and Skrentner, R.G. (1987) Controlling Low Pressure Centrifugal Blowers—A Tutorial. *The Communicator,* **7,** 2.

Manross, R.C. (1983) *Wastewater Treatment Plant Instrumentation Handbook.* EPA-600/8-85-026, U.S. EPA, Cincinnati, Ohio.

National Council of the Paper Industry for Air and Stream Improvement (NCASI) (1984) *A Study of Recently Developed Continuous Dissolved Oxygen Measurement Systems Based on Their Field Performance.* Technical Bulletin No. 440, New York, N.Y.

Nisenfeld, A.E. (1982) *Centrifugal Compressors: Principles of Operation and Control.* Instrum. Soc. Am., Research Triangle Park, N.C.

North American Water and Wastewater Instrumentation Testing Association (ITA) (1988) *A Collection of Seven Reports on Individual On-Line DO Meter Performance.* Washington, D.C.

Novotny, V., *et al.* (1991) Time Series Analysis Models of Activated Sludge Plants. *Water Sci. Technol.*, **23**, 1107.

Ogata, K. (1970) *Modern Control Engineering.* Prentice-Hall, Englewood Cliffs, N.J.

Olsson, G., and Andrews, J.F. (1981) Dissolved Oxygen Control in the Activated Sludge Process. *Water Sci. Technol.*, **13**, 341.

Olsson, G., *et al.* (1985) Self Tuning Control of the Dissolved Oxygen Concentration in Activated Sludge Systems. In *Instrumentation and Control of Water and Wastewater Treatment and Transport Systems.* Int. Assoc. Water Pollut. Res. Control, Pergamon Press, Oxford, Eng.

Perkins, D.W., Jr., (1990) What Makes Your Treatment Plant Tick? *Operations Forum*, **7**, 15.

Speirs, G.W., and Hill, R.D. (1987) Field Verification of On-line Instrumentation at a Municipal Wastewater Treatment Plant. *Water Sci. Technol.*, **19**, 669.

Stenstrom, M.K., and Andrews, J.F. (1980) Cost Interactions in Activated Sludge Systems. Proc. Paper 15267, Am. Soc. Civ. Eng., *J. Environ. Eng. Div.*, **106**, EE4.

Stephenson, J.P. (1985) Practices in Activated Sludge Process Control. In *Comprehensive Biotechnology: The Principles, Applications and Regulations of Biotechnology in Industry, Agriculture and Medicine.* Moo-Young, M. (Ed.), Pergamon Press, Oxford, Eng., **4**, 1131.

Thompson, D., *et al.* (1989) Step Feed Control to Minimize Solids Loss During Storm Flows. *J. Water Pollut. Control Fed.*, **61**, 1658.

Thompson, D., *et al.* (1992a) The Feasibility of Implementing Step Feed Control of Storm Flow at Selected WPCPs: Phase I—Preliminary Investigation. Report prepared for Ont. Ministry Environ., Water Resour. Branch, Toronto, Ont., Canada.

Thompson, D., *et al.* (1992b) Step Feed Control of Storm Flow: Full Scale Implementation. Paper presented at the 65th Ann. Conf. Water Poll. Control Fed., New Orleans, La.

Torpey, W. (1948) Practical Results of Step Aeration. *Sew. Works J.*, **20**, 781.

Vaccari, D. (1986) Solids Flow and Distribution in the Step-Feed Activated Sludge Process. *Civil Eng. Pract. Des. Eng.*, **5**, 877.

Vaccari, D.A., and Christodoulatos, C. (1989) Comparison of Several Control Algorithms for Activated Sludge Waste Rate. *Water Sci. Technol.*, **21**, 1249.

Vaccari, D.A., and Christodoulatos, C. (1990) Estimation of the Monod Model Coefficients for Dynamic Systems Using Actual Activated Sludge Plant Data. *Proc. 5th IAWPRC Workshop*, Kyoto, Jap., 449.

Vaccari, D.A., *et al.* (1985) Calculation of Mean Cell Residence Time for Unsteady-State Activated Sludge Systems. *Biotechnol. Bioeng.*, **27**, 695.

Vaccari, D.A., *et al.* (1988) Feedback Control of Activated Sludge Waste Rate. *J. Water Pollut. Control Fed.*, **60**, 1979.

Wadsworth, H.M. (1990) *Handbook of Statistical Methods for Engineers and Scientists*. McGraw-Hill, Inc., New York, N.Y.

Water Environment Federation (1993) *Instrumentation in Wastewater Treatment Facilities*. Manual of Practice No. 21, Alexandria, Va.

Water Pollution Control Federation (1988) *Aeration*. Manual of Practice No. FD-13, Washington, D.C.

Wesner, G.M., *et al.* (1977) *Energy Conservation in Municipal Wastewater Treatment*. EPA-430/9-77-011, U.S. EPA, Washington, D.C.

Chapter 10
Sequencing Batch
Reactors

*I*NTRODUCTION

The term *sequencing batch reactor* (SBR) is a relatively recent term that has developed into general usage since approximately 1979 (Transfield, Inc.). It refers to a generic system of variable-volume activated-sludge treatment in which aeration, sedimentation, and effluent decant are sequential operations within a single reactor. Consequently, there are no dedicated secondary clarifiers or associated facilities to return waste solids. The number of reactors generally depends on the maximum wastewater volume to be treated.

Sequencing batch reactor technology has evolved rapidly in the U.S. during the last decade. The major vendors of SBR equipment each have their own design approaches. Other SBR systems are unique and reflect the preferences of the design engineer. Design approaches often use different tank and flow configurations, and a variety of options for aeration, independent mixing, effluent discharge equipment, and solids wasting. Systems are typically configured to vary their operation automatically in response to changes in influent flow rates or to accommodate operator-initiated changes

---

either in total cycle times or in the cycle times of individual components (for example, amount of aeration or mixing during fill).

PROCESS CONFIGURATIONS

One classification of SBR systems distinguishes those that operate with continuous feed and intermittent discharge (CFID) from those that operate with intermittent feed and intermittent discharge (IFID).

CONTINUOUS INFLUENT SYSTEMS. Continuous feed–intermittent discharge reactors receive influent wastewater during all phases of the treatment cycle. When there is more than one reactor, as is typically the case for municipal systems, the influent flow is split equally to the various reactors on a continuous basis. For two-reactor systems, it is normal to have the reactor cycle operations displaced so that one SBR is aerating while the second SBR is in the settling and decant phases. This makes it possible to aerate both reactors with one blower continuously in operation and also spreads the decant periods so that there is no overlap. The dry weather flow cycle time for most CFID systems is generally 3 to 4 hours. Each cycle typically devotes 50% of the cycle time to aeration, 25% to settling, and 25% to decant. Stormwater flows are accommodated by reducing cycle time. Under extreme flow conditions, the reactor may operate as a primary clarifier (no aeration phase) with the decanters set at top water level (TWL).

With a CFID system, TWL occurs at the start of the decant phase. Because CFID systems generally operate on the basis of preset time cycles, TWL varies for each cycle as a function of the influent flow for that particular cycle. A typical operating cycle for a CFID reactor is presented in Figure 10.1. The actual effluent flow rate during the discharge event depends on the number of reactors and the percentage of each cycle devoted to decant.

A key design consideration with CFID systems is to minimize short-circuiting between influent and effluent. Influent and effluent discharge are typically located at opposite ends of rectangular reactors, with length-to-width ratios of 2:1 to 4:1 being common. Installation of a prereaction chamber separated by a baffle wall from the main reaction chamber is also a standard feature of some systems.

INTERMITTENT INFLUENT SYSTEMS. In the U.S., the IFID types of systems are sometimes referred to as the conventional, or "true," SBR systems. The one common characteristic of all IFID systems is that the influent flow to the reactor is discontinued for some portion of each cycle. A commonly observed implementation of an IFID system for a wastewater treatment plant (WWTP) containing two SBRs is presented in Figure 10.2.

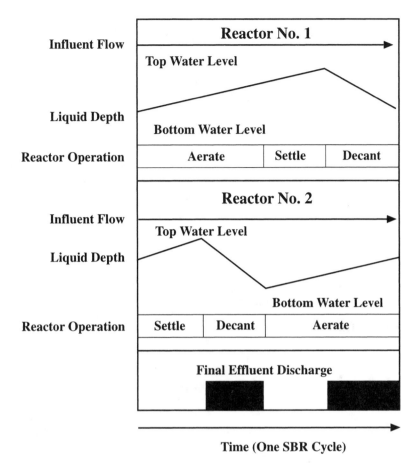

Figure 10.1   Typical operation for a two-reactor continuous feed and intermittent discharge system (SBR = sequencing batch reactor)

Each reactor illustrated in Figure 10.2 operates with five discrete phases during a cycle. During the period of reactor fill, any combination of aeration, mixing, and quiescent filling may be practiced. Mixing independent of aeration can be accomplished by using jet aeration pumps or separate mixers. Some systems distribute the influent over a portion of the reactor bottom so that it will contact settled solids during unaerated and unmixed fill. The end of the fill cycle is controlled either by time (that is, fill for a preset length of time) or by volume (that is, fill until the water level rises a fixed amount). Flow information from the WWTP influent flow measurement or from the rise rate in the reactor determined by a series of floats may be used to control the time allocated to aeration, mixing, or filling in accordance with previously programmed instructions.

At the end of the fill cycle depicted in Figure 10.2, all influent flow to the first reactor is stopped, and flow is diverted to the second reactor. Con-

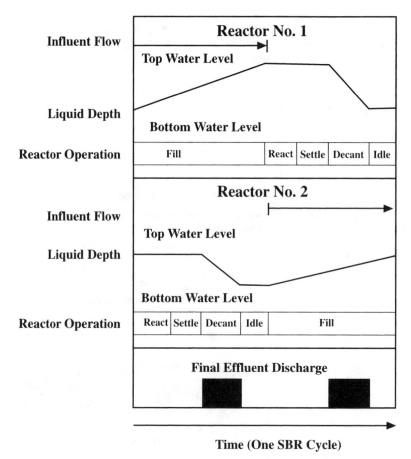

Figure 10.2 **Typical operation for a two-reactor intermittent feed and intermittent discharge system with fill, react, settle, decant, and idle phases (SBR = sequencing batch reactor)**

tinuous aeration occurs during the react phase for a predetermined time period (typically 1 to 3 hours). Again, the time devoted to reaction in any given cycle may automatically be changed as a function of influent flow rate. At the completion of the reaction phase, aeration and any supplemental mixing is stopped, and the mixed liquor is allowed to settle under quiescent conditions (typically 30 to 60 minutes). Next, clarified effluent is decanted until the bottom water level (BWL) is reached. The idle period represents that time period between the end of decant and the time when influent flow is again redirected to a given reactor. During high-flow periods, the time in idle will typically be minimal.

As shown in Figure 10.2, the actual flow rate during discharge has the potential to be several times higher than the influent flow rate. Discharge flow rates are critical design parameters for the downstream hydraulic capacity of sewers (in the case of industrial treatment facilities) or processes such as disinfection or filtration.

Another variation of the IFID approach dispenses with a dedicated reaction phase and initiates the settling cycle at the end of aerated fill. Yet another IFID approach allows influent to enter the reactor at all times except for the decant phase so that normal system operation consists of the following phases: (1) fill-aeration, (2) fill-settling, (3) no fill-decant, and (4) fill-idle; these systems also include an initial selector compartment that operates either at constant or variable volume and serves as a flow splitter in multiple-basin systems. Biomass is directed from the main aeration zone to the selector.

Sequencing batch reactor systems can also be designed for nitrification–denitrification and enhanced biological phosphorus removal. In these cases, the cycle times devoted to such processes as anaerobic fill, anoxic fill, mixed/unmixed fill, aerobic fill, and dedicated reaction depend on the treatment objectives. Mineral addition may also be practiced to achieve effluent objectives more stringent than typical secondary effluent requirements. Systems can also be configured to switch from IFID operation to CFID operation when necessary to accommodate stormwater flows or to allow a basin to be removed from service while still treating the entire WWTP flow in a remaining basin. The one common factor behind all SBRs is that aeration, settling, and decant occur within the same reactor.

SEQUENCING BATCH REACTOR EQUIPMENT

PROCESS CONTROL. The programmable logic controller (PLC) is the optimum tool for SBR control and all present-day vendors use this approach. Sequencing batch reactor manufacturers supply both the PLC and required software. Typically, programs are developed and modified by the SBR vendor using a desk-top computer and software supplied by the PLC vendor. Vendor-developed programs are proprietary and may not be modified by the design engineer or the WWTP operator. Depending on the proprietary software design and type of system, the operator may independently select such variables as solids waste rates; storm cycle times; and aeration, mixing, and idle times. In addition, the design engineer may develop additional software to interface the PLC to a desk-top computer for graphic presentation of process operation to the operator and generation of archive data and compliance reports.

Programmable logic controller hardware is of modular construction. Troubleshooting procedures are well defined, and replacement of a faulty module is not difficult. An internal battery protects the software in the event of power failure. The software is backed up by a memory chip (EPROM) and can be easily reloaded if the battery fails. The PLC expertise required of the owner is limited to maintenance and repair functions that are well within the capability of a competent electrician.

REACTORS. Reactor shapes include rectangular, oval, circular, sloped sidewall, and other unique approaches. Design TWLs and BWLs often allow decanting from 20 to 30% of the reactor contents per cycle.

DECANTERS. Some decanters are mechanically actuated surface skimmers that typically rest above the TWL. The decanter is attached to the discharge pipe by smaller pipes that both support and drain the decanter. The discharge pipe is coupled at each end through seals that allow it to rotate. A screw-type jack attached to a worm gear, sprocket, and chain to an electric motor rotates the decanter from above the TWL to BWL. The speed of rotation is adjustable.

Other decanters are floated on the reactor surface. These decanters may approximate a large-diameter plug valve, whereby the top portion acts as the valve seat (and provides flotation). The bottom is the plug that is connected to a hydraulic operator that moves it away from the seat (approximately 50 mm [2 in.]) to allow discharge, or back to the seat to stop discharge. Other floating decanters consist of a length of pipe suspended on floats, with the pipe having a number of orifices bored in the bottom. The number of orifices (and length of pipe) is flow dependent. Each orifice is blocked by a flapper or plugs to prevent solids entry during aeration. There are also decanter configurations that float an effluent discharge pump.

Other decanters are typically fixed-position siphons located on the reactor wall. The bottom of the decanter (collection end of the siphon) is positioned at the BWL. Flow into the decanter is under a front lip (scum baffle), over an internal dam, and out through a valve. When the water level in the reactor falls below the front lip, air enters the decanter, breaking the siphon and stopping flow. The trapped air prevents mixed liquor from entering during the reaction and settling modes. At the end of settling, the trapped air is released through a solenoid valve and the siphon is started.

SOLIDS WASTING. The wasting of both aerated mixed liquor suspended solids (MLSS) and settled MLSS is practiced. The wasting systems frequently consist of a submersible pump with a single point for withdrawal. Gravity flow waste systems are also used. Another approach uses influent distribution piping for multiple-point withdrawal of the settled solids.

AERATION/MIXING SYSTEMS. A variety of aeration and mixing systems are in use with SBRs. These include jet aeration, fine- and coarse-bubble aeration, and turbine mechanical aeration. Some systems use a floating mixer to provide mixing independent of aeration. Other diffused aeration facilities do not have any mixing capability independent of aeration. Independent mixing is readily obtained with a jet aeration system.

SUMMARY

Sequencing batch reactor technology has gained rapid acceptance in the U.S. within the past 5 years. A variety of approaches with regard to hydraulics, system cycle operation, and equipment is being used to implement SBR technology. Past equipment-related problems appear to be largely overcome, and reliable SBR components are currently available. Operator acceptance of the technology is high (Deeny et al., 1991).

REFERENCES

Transfield, Inc., *Cyclic Activated Sludge System: A Cost Effective Approach to Waste-Water Treatment*. Irvine, Calif.

Deeny, K., *et al.* (1991) Implementation of Sequencing Batch Reactor Technologies in the United States. Paper presented at 64th Annu. Conf. Water Pollut. Control Fed., Toronto, Can.

Chapter 11
Control Strategies for Fixed-Growth Reactors

INTRODUCTION

Trickling filters (TFs) and rotating biological contactors (RBCs) represent the bulk of the fixed-growth reactors (FGRs) used in the treatment of municipal wastewater. There are now FGR designs such as the downflow 'Bicarbone' process (Stensel *et al.*, 1988), upflow fluidized bed reactors, and submerged expanded media reactors like the 'Captor' process (Atkinson *et al.*, 1979) and the 'Linpor' process (Hegeman, 1984). Newer processes using reticulated particles in the aeration basin include 'Biofor' (Pujol *et al.*, 1993), 'BioStyr' (Toettrup *et al.*, 1993), and 'KMT MBBR' (Ødegaard *et al.*, 1993). These newer processes may be either aerobic or anaerobic. The process control strategies for these newer and developing technologies are beyond the scope of this discussion, and the references, recent literature, and suppliers should be contacted.

This discussion is limited to the presentation of control strategies that will optimize the performance of TFs and RBCs and eliminate operating prob-

---

lems, both general and site specific. Both of these types of FGRs have been considered relatively uncomplicated unit processes that do not require significant operator attention. This perception of TFs has blocked the use of knowledge gained many years ago that would have enhanced their performance and furthered the acceptance of these units for advanced wastewater treatment.

TRICKLING FILTERS

The TF process (WEF, 1992) consists of a fixed-film support material—rock, plastic-form particles, polyvinyl chloride (PVC) bundles, or wood lath—that has been stacked in a container. The media are placed on a bottom structure or false floor, which allows air to flow under and through the media stack and the effluent to be collected for removal. The wastewater is discharged to the top of the media stack by a rotating distributor that intermittently doses the TF media. The applied flow percolates (trickles) down through the media-contacting biomass growing on the media. The biomass removes the organic matter and is maintained by sloughing of the excess growth. Effluent is collected from the underdrain area; a portion may be directly recycled, and the remainder flows to further biological treatment or to clarification.

Process control strategies should be directed to those aspects of TF design and operation (Figure 11.1) that can enhance performance, reduce

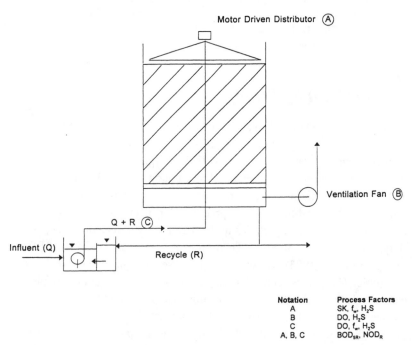

Notation	Process Factors
A	SK, f_w, H_2S
B	DO, H_2S
C	DO, f_w, H_2S
A, B, C	BOD_{SR}, NOD_R

Figure 11.1   Control functions in a trickling filter

operating deficiencies, or eliminate unaesthetic conditions that detract from the acceptance of the process. It is often possible that TFs that are considered to be performing well can perform better or operate at higher capacity and be relatively odor free. In general, process control strategies for TFs should ensure that

- Media wetting efficiency is high,
- Liquid flow distribution is uniform,
- Excess biomass buildup is minimized,
- Filter media is well ventilated to maintain an aerobic biomass,
- Excessive winter cooling is prevented,
- Predator growths are controlled,
- In-depth media fouling is prevented, and
- Odor levels are minimal.

The techniques to ensure that these functions are optimized are discussed in Water Environment Federation Manual of Practice No. 8, *Design of Municipal Wastewater Treatment Plants* (1992).

MAINTENANCE CONTROL STRATEGIES. Maintenance strategies include routine maintenance of the uniform flow distribution system. This would include periodic pan testing to ensure that a uniform discharge of flow is being applied to the surface of the media. Secondly, the distribution nozzles must be routinely checked and any blockage removed. The number of arms in operation should be minimized to increase distribution efficiency and reduce windage cooling.

There is an increasing amount of matter that is nonbiodegradable synthetic material (trash) appearing in primary effluents. These materials lodge in the upper portions of the media and cause localized obstruction of flow and an opportunity for excess biomass to collect, anaerobically decompose, cause odors, produce nuisance organisms, and otherwise detract from the operation and performance of the TF. If there is only coarse screening of the influent, it may be necessary to prescreen primary effluent using 3- to 6-mm climber or rotating screens or place a screen that is 6 mm or finer on the media surface to trap this trash material.

Excessive cooling of trickling filters may be a design, maintenance, and/or operating problem. In cold climates, it is necessary to have wind barriers that extend 1.5 to 2.5 m above the media to reduce heat losses. Further, it is necessary to reduce recirculation and control air flow to limit temperature reduction. For domestic wastewater, a temperature loss of 1 to 1.5°C across the TF virtually ensures that there is sufficient air flow unless windage thermal losses above the media are high.

OPERATING CONTROL STRATEGIES. The optimum wetting efficiency, biomass control, maintenance of an aerobic biomass, and control of odors are important and interrelated aspects of the operation of TFs. *Design of Municipal Wastewater Treatment Plants* (WEF, 1992) provides detailed information regarding the process and performance interrelationships.

Differences in the liquid and ambient air temperatures are typically adequate to draft the required air quantities naturally. Unfortunately, the differences may not be sufficient during critical loading periods of the day and during the spring and fall. The lack of adequate ventilation results in odors and reduced performance, often accompanied by sloughing events. Power ventilation has been recommended for all TFs employing rock and plastic media (WEF, 1992). The lack of adequate air flow cannot be compensated for by other measures.

The fans providing the forced ventilation may be operated in either upflow or downflow mode. Odors will be minimized by using the downflow mode because the lower reaches of the TF media can remove odorous organic compounds, and upflowing air can strip volatile organics from the influent wastewater.

The operation of the fans would be continuous in most applications because the power requirements are low, less than 1 kW/235 L/s (0.25 hp/mgd). However, the design could incorporate the use of additional automatically controlled fans that are on line during peak loading periods.

Air flow distribution is more difficult when the forced ventilation counterflows natural draft. Localized upflow vapor plumes in winter with downflow forced air are evidence that air distribution is not optimal. Better distribution can be achieved by forcing air in the upflow mode through the media during this period. This would require reversible axial flow fans for the air supply. However, ventilation may be less critical during cold periods, and the counterflow to downflow typically may be ignored.

Forced ventilation is required when the density difference of air inside and outside the TF is less than that required for adequate natural draft. Because of the humidity effects on density, temperature differences inside and outside the TF are not an absolute measure for defining when natural draft is sufficient. However, a difference of 5°C between inlet water and ambient air will typically provide adequate natural draft with a well-designed natural ventilation system. Therefore, an automatically controlled draft design could be provided for larger installations, where power savings may be considered worthwhile.

The control mechanisms for the ventilation fans may take three forms:

- Continuous air flow 24 hours per day,
- Modulated air flow as a function of estimated organic loading (seasonal or partial-day industrial loads), or

- Part-time power ventilation based on either air density or temperature differences.

The dissolved oxygen (DO) in the discharged effluent is not a good indicator of adequate ventilation. The treated effluent falls from the media to the floor and can be reaerated from 0 to 3–5 mg/L DO. Furthermore, the loading and oxygen demand are much lower in the bottom of the TF, and deficient zones are typically in the upper 30 to 40% of the media volume.

POWERED FLOW DISTRIBUTORS

Electrically powered distributors can enhance the performance and reduce TF odors, whether employing rock or plastic media (Albertson, 1995, and WEF, 1992). The rationale and the basis of design for the distributor operation are provided in these publications.

The hydraulic term used to define the dosing effects of modifying the distributor speed is Spülkraft, or SK, which is determined from the expression:

$$\text{SK, mm/pass (of an arm)} = \frac{(q + r, \text{ m}^3/\text{m}^2 \times \text{h})(1\,000 \text{ mm/m})}{(a, \text{ no. of arms})(n, \text{ rev/min})(60 \text{ min/h})} \quad (11.1)$$

Where

$q + r$ = total hydraulic application rate.

In effect, the SK is defined by the depth of water (mm) resulting from the pass of one arm of the distributor over a pan.

The distributor speed control employs three different concepts. They are

1. A two-speed arrangement, in which the speed is slowed during low-flow loading periods (12:00 to 5:00 a.m.) to flush the media. The distributor speed is adjusted to the optimum SK for the balance of the operating day.
2. A two-speed arrangement for constant flow and loading of industrial applications, in which the low-speed flushing cycle is repeated several times per day, for example, 20 minutes flushing SK and 3 hours and 40 minutes optimum SK on a 4-hour repetitive cycle.
3. A modulated SK as an inverse function of the loading or influent flow to effect flushing and distribute the organic loading to the full depth of the biotower.

The simplest two-speed control arrangement is shown in Figure 11.2. The same control output can be developed by a microprocessor to control

TRICKLING FILTER A

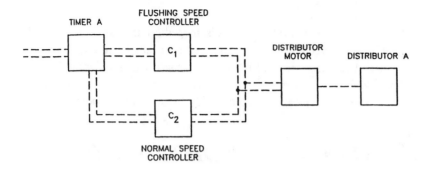

TRICKLING FILTER B

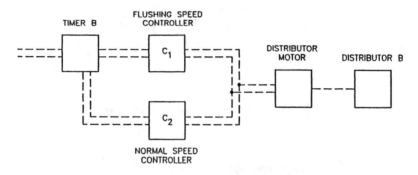

TIMER A & B – 24 HOUR REPEAT TIMERS

Figure 11.2   Control and drive arrangement for two-speed distributor (WEF, 1992)

variable-speed drives for the distributor. The drive could be an electric or hydraulic motor with a 15:1 to 20:1 speed range installed at the center or periphery of the trickling filter. Individual drives for each TF permit the simultaneous evaluation of different operating conditions to establish optimum SK ranges.

The two-speed drives would control the distributor operation to produce the SK at the flow (load) conditions set forth in Figure 11.3. The length of the flushing cycle should be adjustable (Figure 11.3, top) and the time/cycle and cycles/day adjustable in the alternative mode (Figure 11.3, bottom), which is used for heavily loaded roughing towers or industrial biofilters.

The SK value could be continuously modulated as a function of flow. In this manner, the retention time in filter media could be controlled in a manner to develop biomass growth throughout the media depth rather than

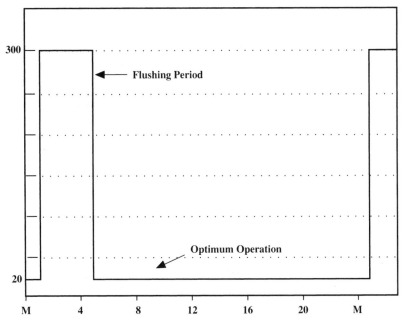

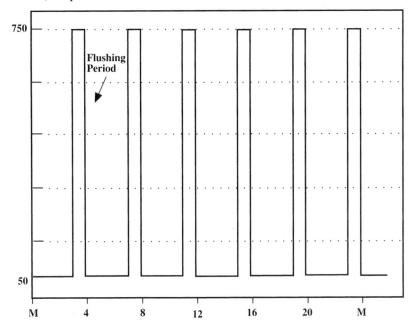

Figure 11.3 Two-speed operation for media flushing: top, municipal operating Spülkraft (SK) cycle; and bottom, industrial operating SK cycle

---

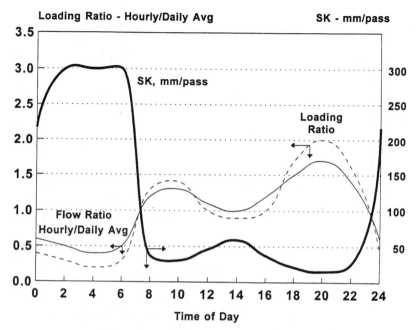

Figure 11.4   Modulated Spülkraft (SK) from flow or load control

mostly in the upper region (Albertson, 1995). The SK is then increased as a function of flow and loading to enhance overall performance, with an operational characteristic as shown in Figure 11.4. The formula (Albertson, 1995) used for the algorithm is

$$\text{SK} = L \, (Q/Q_A)^{-X} \qquad (11.2)$$

Where

L =   site/loading specific constant, mm/pass;
Q =   instantaneous flow including recycle, L/s;
Q_A =   average flow including recycle, L/s; and
X =   site-specific constant, 1.5 to 3.0.

The maximum SK required for flushing is a function of the 5-day biochemical oxygen demand (BOD_5) loading, kg/m^3·d (L_B), as is

$$\text{Flushing SK} = 240 + 125 \, L_B \qquad (11.3)$$

With an operable speed range of 15:1 (with 20:1 VFD) the minimum SK is

$$\text{Minimum SK} = \frac{\text{Flushing SK}}{15} \qquad (11.4)$$

and the exponent X would be defined by:

$$\left(\frac{Q_{min}}{Q_{max}}\right)^X = 15 \qquad (11.5)$$

The modulating SK will require either a flow signal input or an algorithm for the drive controller. The signal may be generated by a microprocessor or by an electric drive that has an integral capability to be programmed for several setpoints and ramp speeds between the setpoints. This drive will be able to approximate a municipal wastewater treatment plant diurnal loading or flow curve. The fully modulating SK as a function of flow or load is shown in Figure 11.5, top, and a multispeed drive approximating the diurnal flow load rate is shown in Figure 11.5, bottom.

The distributor drive output range should be 15:1 to 20:1, although operating experience may dictate a smaller minimum-to-maximum SK ratio as indicated in Figure 11.6 (see also Albertson, 1995, and WEF, 1992).

ROTATING BIOLOGICAL CONTACTORS

The rotating biological contactor (RBC) system consists of circular plastic disks mounted on a horizontal shaft in a basin. Typically, approximately 40% of the media is immersed in the wastewater. The shafts are rotated (1 to 2 rev/m) by either a mechanical or compressed-air drive. The shafts are arranged to provide flow through a series of stages in a train containing one to three shafts. The wastewater being treated flows through the contactor by simple displacement and gravity. The rotation of disks alternately exposes the biofilm to the organic material in the wastewater and to atmospheric air. Bacteria and other microorganisms that are naturally present in wastewater adhere to and grow on the surface of rotating media.

Depending on organic loading conditions, each succeeding stage will show varying slime thickness and color. Generally, the first stage shows characteristic brownish-gray color when operated within the reasonable loading range; the last (nitrifying) stages show a reddish-bronze color. Because of shearing forces, the biological film tends to slough off whenever the biomass growth on the media surface becomes too heavy. The sloughed biofilm and other suspended solids are carried away in the wastewater and are removed in the secondary clarifier.

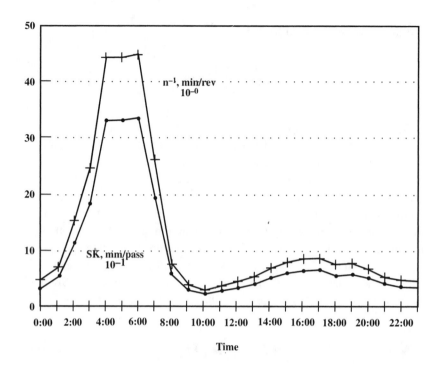

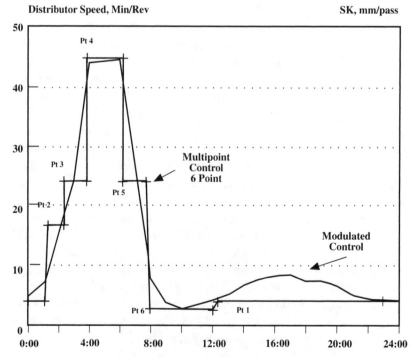

Figure 11.5 Types of modulated Spülkraft (SK) control: top, fully modulated SK control; and bottom, stepped function SK control

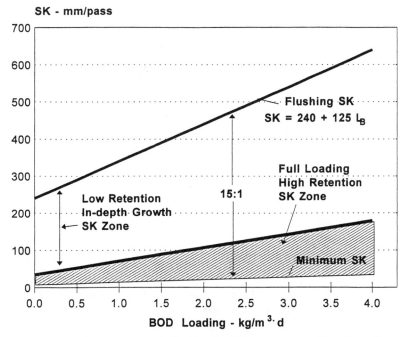

Figure 11.6   Recommended Spülkraft (SK) range as a function of biochemical oxygen demand$_5$ (BOD) loading

DISSOLVED OXYGEN

The aspects of RBC process control center on two fundamental goals: maintaining aerobic conditions and controlling biofilm inventory. The goals overlap for operational control procedures, as shown in Table 11.1.

Recent experiences have favored air-driven drives or supplemental air to mechanical drives over the original mechanical drive designs. The air not only reduces low DO zones, but is more effective in controlling excess biomass buildup. Excessive biomass accumulation and inadequate DO have been the source of many RBC operating problems (Heidman and Gilbert, 1984; Gross *et al.*, 1984; and WEF, 1992). Benefits of submerged aeration and assisted biomass control may outweigh the higher power requirements for the air or combined mechanical/air drive.

AIR DRIVES. The critical DO zones are the first, and sometimes the second, stage of the train. The DO probe in the initial stages can provide the necessary signal to control the air flow automatically to the separate trains or to the total RBC facility. With good flow balancing, the DO probes located in the two or three first compartments would suffice for automatic (or manual) air flow control.

Table 11.1   Control aspects of mechanical and air-driven rotating biological contactors

Item	Dissolved oxygen	Biomass inventory control
Air rate metering	X	X
Rotational speed control	X	X
Step feed	X	X
Secondary effluent recycle	X	
Adjustable staging baffles	X	X
Reverse flow mode	X	X
Biomass stripping	X	X

MECHANICAL DRIVES. While it has been reported (WPCF, 1987) that variable-speed drives can enhance oxygen transfer, specific data and the range of control of DO are not defined. An increase in rotational speed would help control excess biofilm, with the resulting higher shear forces. Other factors that can improve oxygen transfer include supplemental aeration, continuous or intermittent effluent recycle, step feed of the influent, and adjustable baffles to change the volume in the initial stages (WEF, 1992).

Step feeding, effluent recycling, and repositioning of the staging influent baffles all spread (balance) the loading to reduce oxygen deficiency in the initial stages of treatment. The improvement in liquor quality (lower soluble chemical oxygen demand) effected by these procedures would also enhance oxygen transfer through the liquid film. Also, these procedures would serve to limit film thickness, as film thickness is a function of the applied loading.

Alternating the flow path (see Figure 11.7, bottom) (as in the alternating-double-filtration mode for trickling filters) would enhance oxygen transfer, limit biomass buildup, and help control predation. However, it is necessary that standard media be used in all staging and that the staging be established for the flow regime being employed.

To eliminate oxygen-deficient periods, continuous DO monitoring of the initial stages is needed to define limiting DO periods and to establish operating procedures, physical changes, and automatic response.

BIOMASS INVENTORY CONTROL. Design and operational modes that enhance oxygen transfer are generally beneficial to inventory control. Intermittent and continuous aeration is more effective than mechanical drives that rely solely on rotational speed. Reversing the flow path can also be effective. The mode of operation that limits biomass buildup would generally provide the best performance.

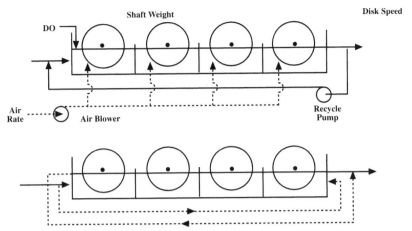

Figure 11.7 Control functions in rotating biological contactor systems: top, conventional flow with effluent recycle option (all modes of disk drive); and bottom, flow reversal for biomass control (all modes of disk drive) (DO = dissolved oxygen)

In some instances, excess biomass will still build up and must be eliminated by direct methods. These methods could include hosing and chemical stripping using a caustic solution. The biomass buildup in the media should be monitored, and load cells can be installed under the bearing blocks for this purpose. The signal can be processed manually or electrically/mechanically for the appropriate response.

Like the TF, the RBC, once established in a stable environment with appropriate loadings and staging, and with adequate DO and control of the biomass, does not require significant operator attention. Therefore, control functions must be directed at monitoring the DO and stage weight and at the implementation of either automatic or manual corrective actions.

REFERENCES

Albertson, O.E. (1995) Excess Biofilm Control by Distributor Speed Modulation. *J. Environ. Eng.*, **121**, 4, 37.

Atkinson, B., *et al.* (1979) Biological Particles of Given Size, Shape and Density for Use in Biological Reactors. *Biotechnol. Bioeng.*, **21**, 2, 193.

Hegeman, W. (1984) A Combination of the Activated Sludge Process with Fixed-Film Bio-Mass to Increase the Capacity of Wastewater Treatment Plants. *Water Sci. Technol.*, **16**, 119.

Heidman, J.A., and Gilbert, W.C. (1984) *Summary of Design Information on Rotating Biological Contactors.* EPA-430/9-84-008, U.S. EPA, Washington, D.C.

Ødegaard, H., *et al.* (1993) A New Moving Bed Biofilm Reactor—Application and Results. *Proc. 2nd Int. Conf. Biofilm React., Int. Assoc. Water Qual.*, Paris, Fr.

Pujol, R., *et al.* (1993) Les Biofiltres: Des Reacteurs Biologiques Adaptables at Fiables. *Proc. 2nd Int. Conf. Biofilm React., Int. Assoc. Water Qual.*, Paris, Fr.

Gross, C., *et al.* (1984) RBCs Reach Maturity. *Water Eng. Manage.*, 28.

Stensel, H.D., *et al.* (1988) Biological Aerated Filter Evaluation. *J. Environ. Eng.*, **114**, 1352.

Toettrup, H., *et al.* (1993) The Treatment Trilogy of Floating Filters. *Proc. 2nd Int. Conf. Biofilm React., Int. Assoc. Water Qual.*, Paris, Fr.

Water Environment Federation (1992) *Design of Municipal Wastewater Treatment Plants*. Manual of Practice No. 8, Alexandria, Va.

Water Pollution Control Federation (1987) *Operation and Maintenance of Trickling Filters, RBCs, and Related Processes*. Manual of Practice No. OM-11, Alexandria, Va.

Chapter 12
Stabilization

AUTOTHERMAL THERMOPHILIC AEROBIC DIGESTION

Autothermal thermophilic aerobic digestion (ATAD) systems are normally two-stage aerobic digestion processes that operate under thermophilic temperature conditions (40 to 80°C) without supplemental heat. Like composting, the process relies on the conservation of heat released during digestion to attain and sustain the desired operating temperature. Typical ATAD systems operate at 55°C and may reach temperatures of 60 to 65°C in the second-stage reactor. Perhaps because of the high operating temperature, this process has been referred to as *liquid composting*.

Figure 12.1 illustrates a typical configuration for ATAD systems that includes prethickening, two enclosed and insulated reactors (configured in series), and a final storage/postthickening tank. Several versions of the technology exist that may use different aeration, mixing, and foam control equipment and may also differ by the feeding strategy used. Most ATAD systems are used in community treatment systems that practice some form

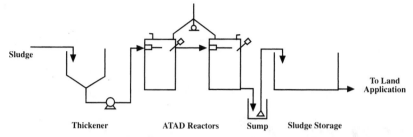

Figure 12.1   Typical autothermal thermophilic aerobic digestion (ATAD) configuration

of beneficial reuse of biosolids and are operated as batch systems to maximize pathogen reduction (U.S. EPA, 1990).

In a two-reactor ATAD system, a total treatment time of 5 to 6 days (2.5 to 3 days per reactor) is adequate to satisfy the process requirements for destruction of pathogens and total organic solids. For ATAD systems treating a mixture of primary and waste activated sludges, volatile solids reduction is expected to be within the range of 35 to 45%.

CONTROL STRATEGIES. Autothermal thermophilic aerobic digestion systems are not heavily instrumented, nor are they mechanically complex. Many of the early ATAD systems were operated manually without the need for significant automation. More recent installations include monitoring instrumentation for temperature (two probes per reactor), suspended solids concentration, feed volume, and pH. Newer installations can also be equipped to control automatically the draw-and-fill cycle for the multiple-stage system. This control is accomplished through the use of a programmable logic controller (PLC), automatic valve operators, and reactor level sensors.

Most operators prefer to run the ATAD process manually or semiautomatically, even if the system is equipped with the necessary equipment for full automatic control. Semiautomatic operation means that the program for withdrawal and filling is started manually.

Process control has three major focuses: (1) temperature, solids, and pH monitoring; (2) feed sludge control; and (3) control of the draw-and-fill cycle. In most cases, process control consists of performing periodic suspended solids and pH analyses, monitoring reactor temperatures, and controlling the pumping of specific volumes of solids to the ATAD reactors on a batch basis.

PATHOGEN REDUCTION CONTROL. Autothermal thermophilic aerobic digestion is recognized as a Process to Further Reduce Pathogens (PFRP) and, as such, is considered capable of reducing pathogens to levels defined by the U.S. Environmental Protection Agency (U.S. EPA) for

"Class A sludge," called biosolids (U.S. CFR, 1993). Regulations that define the criteria for treated solids also define operating conditions that are required to be met during operation (U.S. CFR, 1993). Solids are required to be treated for a specific time period and at a specified temperature to achieve the level of pathogen reduction performance required for least restricted beneficial use of the treated product. Figure 12.2 illustrates the defined time–temperature conditions necessary to achieve the desired pathogen reduction performance goals. For those facilities that operate ATAD facilities with pathogen reduction goals, this time–temperature relationship becomes an important focus of process control.

TREATMENT TIME CONTROL. Most ATAD systems are operated as batch processes that receive sludge once each day. The sludge volume that is fed to the system each day amounts to approximately one-sixth of the total effective tank volume, which results in a hydraulic residence time of 6 days. Although the concept of residence time is adequate to define the requirement for volatile solids reduction, it is not adequate for pathogen control because in a completely mixed environment, the tank contents represent a mixture of suspended matter (and pathogens) with varying degrees of retention times. The minimum retention time is the period of time between feed events during which the digestion tanks are isolated.

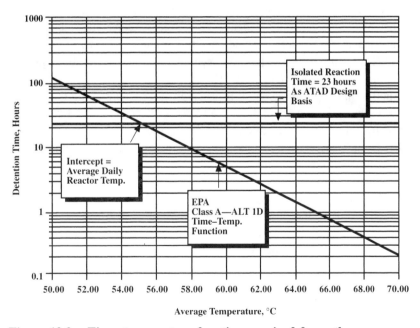

Figure 12.2   Time–temperature function required for pathogen reduction (ATAD = autothermal thermophilic aerobic digestion)

Stabilization   165

Feed pumps for most ATAD systems are designed to deliver the required thickened sludge volume to the reactor(s) in less than 1 hour each day. The reactors remain isolated for the remaining 23 hours each day. This time period (23 hours) is the minimum retention time for suspended matter and pathogens in the system. As indicated in Figure 12.2, a minimum average temperature of 55°C is required for pathogen control in a system with a 23-hour retention time. The strategy for controlling the retention time of the process is simply to control the duration of the draw-and-fill cycle.

TEMPERATURE MONITORING AND CONTROL. Autothermal thermophilic aerobic digestion systems are by definition self-heating. Maintenance of optimum temperature is facilitated by efficiently conserving the heat released during digestion. Solids that are fed to the system are the "fuel" that drives the process. The mass of feed to the system is directly related to the temperature that is observed in the process. Because the volume of solids fed to the system is, for the most part, a constant, the raw volatile solids concentration is the main process variable that influences operating temperature. A minimum volatile solids concentration of 2.5% is typically sufficient to maintain the minimum average second-stage reactor temperature at or greater than 55°C (U.S. EPA, 1990). Many operating facilities are known to achieve temperatures to 65°C. Although pathogen destruction and volatile solids reduction may be enhanced at the elevated temperatures, it is generally recognized that process temperatures should be maintained at 65°C or less to avoid operating problems (Deeny *et al.*, 1991). This will typically require the feed volatile solids concentration to be maintained at or at 5% or less.

Short-term temperature control is accomplished by controlling the feed solids to the system. A more dilute feed lowers temperatures, and a higher concentration raises temperatures. Over the long term, if temperature control is problematic, it points to the need for a heat exchanger to either recover or remove heat from the system.

DRAW-AND-FILL CYCLE. The largest number of ATAD systems are operated in a draw-and-fill mode to enhance pathogen destruction. The filling process is typically initiated by the wastewater treatment plant (WWTP) operator. The process begins by shutting down the aerators and removing approximately one-third of the second-stage reactor volume. The two reactors are then allowed to equalize, and influent sludge is pumped into the bottom of the first reactor until warm sludge is displaced to the second stage to fill the reactor volume. Once at full level again, the aerators are reactivated and the reactors are aerated in complete isolation until the operator (or the PLC) initiates feeding again—in approximately 23 hours.

MONITORING MIXING CONDITIONS. Mixing and temperature conditions are important design and operating issues for ATAD systems. Maintaining adequate mixing conditions, and thereby maintaining uniform reactor temperature conditions, is an important aspect of process control. Two thermocouples are installed in each ATAD reactor at opposing locations to monitor temperature conditions. Their output is typically directed to a strip chart recorder for monitoring. Under adequate mixing conditions, both temperature trends are in general agreement and coincide. However, the trends diverge when adequate mixing is not occurring, such as during feed events.

Mixing requirements are established by design and are not generally subject to a control strategy. However, adequate mixing is extremely important for ATAD systems, given the rate at which treatment reactions are taking place. Operational monitoring is required to detect impaired mixing conditions that can result from rag accumulation on aerator shafts, plugged nozzles, impeller wear, or similar circumstances.

FOAM CONTROL. Foam generated during the digestion process forms a floating layer on the surface of the sludge in the reactor. Although foam is considered a nuisance in most digestion systems, it performs important functions in ATAD systems. The foam layer is believed to improve oxygen use, to contribute to biological activity that occurs in the foam layer itself, and to insulate the liquid surface (Deeny *et al.*, 1991). For these reasons, foam is managed but not eliminated from the reactors.

Foam control is established by design through the installation of control devices in the reactor above the liquid surface. These devices impinge the foam surface and break down the size of the foam bubbles. This results in the formation of a dense foam layer with good insulation properties.

ODOR CONTROL. As a result of the high operating temperature, nitrification is suppressed during digestion. Therefore, nitrogen release that occurs as a result of digestion remains in the form of ammonia, which is detectable in the reactor exhaust. Odor control measures, when required, typically include biofilters to treat the exhaust gas. A typical system is illustrated in Figure 12.3. The system consists of a simple rectangular tank(s) containing a bed of compost, bark, or similar material. Microbes that adhere to the surface of the bed material treat odor-producing compounds as the humid air stream passes through the bed. A bed moisture content between 40 and 60% is necessary to sustain the system's biological culture. For this reason, biofilters are preceded by a spray chamber (or similar device) to humidify the air input to the biofilter.

ANALYTICAL MONITORING. Periodic analyses for total solids, suspended solids, total volatile solids, and volatile suspended solids are required to monitor feed conditions and digestion performance. U.S. EPA

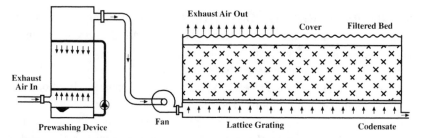

Figure 12.3   Typical biofilter used for exhaust air treatment (U.S. EPA, 1990)

has presented, with illustrative examples, several methods for determining volatile solids reduction by digestion (U.S. EPA, 1992). One of these methods should be adopted and used for process control. Measurements of pH can also be used as a secondary indication of digestion progress, but pH is not recommended for use as the basis of control.

*A*NAEROBIC DIGESTION CONTROL

Anaerobic digestion is used to decompose the solids produced and removed in wastewater treatment processes. The objective of the digestion process is to reduce the mass and volume of the solids while generating potentially useful byproducts and destroying or controlling agents of disease and infection.

The anaerobic digestion process is a multistage biochemical process in which organic compounds are broken down to methane, carbon dioxide, and water. In the first stage, cellular enzymes reduce solid complex organic compounds (cellulose, proteins, lignins, and lipids) into soluble organic fatty acids, alcohols, carbon dioxide, and ammonia. These products are then converted by acid-forming bacteria into acetic acid, propionic acid, hydrogen, carbon dioxide, and other low-molecular-weight organic acids. In turn, methane-forming bacteria convert the hydrogen and carbon dioxide to methane and convert the acetates to methane and bicarbonate. All three steps or stages occur simultaneously in a given digester, and successful operation requires a balance between the production and consumption of products of each stage.

The methane-forming bacteria exhibit the slowest waste conversion rate under normal operating conditions. Therefore, to achieve high levels of stabilization, it is necessary to monitor certain biological and physical factors that affect these methane-forming microorganisms; factors to be monitored are summarized in Table 12.1.

Of these physical and chemical factors, only temperature, flow, solids concentration, and the mixing functions can be automatically controlled

Table 12.1   Factors in anaerobic digestion

Physical factors	Chemical factors
Temperature	Alkalinity
Sludge feed and withdrawal, and gas flow rates	Volatile acids
Hydraulic retention time	Gas methane content
Solids retention time	Nutrients
Solids loading	Trace elements
Mixing	Toxic compounds
Solids concentration	
Sludge type	
Volatile solids loading	

and/or monitored. The other information is derived from laboratory analyses and calculations using physical dimensions of the digester, manually entered values, and on-line measurements. The resulting information is analyzed by the operations personnel to monitor digester "health" and ensure efficient digester operation.

DIGESTER LOADING. As with any biological process, it is most desirable to have a continuous flow of nutrients to the process; however, in loading a digester, the operator does not typically have that luxury, but must load when feed sludge (sludge from the clarification process) is available. The operator can, if there are storage or thickener facilities before the digestion process, manipulate the loading by adjusting the feeding cycle and varying the feed solids concentration.

In feeding a primary digester, the operator must be able to monitor and control the influent flow rate. This is accomplished by using a flow meter, a flow controller, a totalizer, and a modulating valve or variable-speed pump. The operator establishes the desired flow rate and totalized flow setpoint to be fed each cycle by entering the desired values at the control station. These values are generally established to provide the digester with at least one feeding per hour. Feedback is provided at the valve so that the operator may reevaluate the flow rate settings if the valve is being excessively throttled (Figure 12.4).

When there is more than one primary digester in the treatment process, the digesters may be fed sequentially using two-position control valves on their influent lines. Valve status (opened/closed) is monitored to indicate which digester is being fed. This information is used by the flow totalizer to calculate the sum of influent flow to the digester while the associated influent valve is open. The sequencing controller sequentially operates the influent valves in accordance with the totalized digester flow and the cycle totalized flow setpoint. The controller opens the next-sequenced valve before closing the valve currently open. The accumulated flow to each

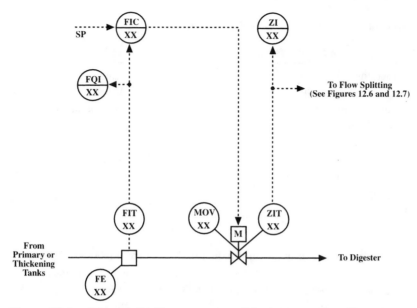

Figure 12.4   Digester feeding process and instrumentation diagram

digester is maintained on a daily basis, along with the total flow to the digester battery. The accumulated flow to each digester is displayed for the operator's use (Figure 12.5).

In larger WWTPs with multiple batteries of digesters (typically four to six digesters per battery), the batteries are fed simultaneously with one digester in each battery being loaded at a time. Care must be executed in the control strategy to ensure that a valve is always open in the event that positive displacement pumps are used to feed the digester. A closed valve could cause significant damage to the pump and/or piping system. Each battery is provided with a flow meter, a controller, and a modulating control valve to establish the flow to each battery. The operating parameters within each battery are maintained as previously described. To ensure that the flow is appropriately split between the batteries, two types of control algorithms, *master/slave* and *most-open valve*, are available to establish the flow rate setpoints for each battery. Both algorithms use the feed cycle setpoint values for each digester to establish the flow rate setpoints for each battery.

In the master/slave algorithm, one battery is established as the master, and the associated inlet valve is automatically positioned at some operator-adjusted value (90% open). The flow setpoints for the slave battery, or batteries, are a ratio of the measured master battery flow. This ratio is the value determined by dividing the totalized feed setpoint values for each digester in the slave battery by the total for the master battery (Figure 12.6).

In the most-open valve algorithm, an additional controller is required to control the most-open valve to an operator-established setpoint. Valve position is monitored for each battery-feed valve, and the most open is selected

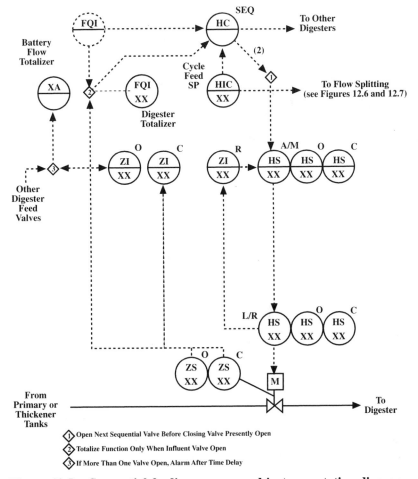

Figure 12.5   Sequential feeding process and instrumentation diagram

as the input to the valve position controller. The output from the valve position controller represents the *total flow setpoint* for the digestion system. This value is cascaded to the individual battery flow controllers, where it is multiplied by a gain factor and used as the individual battery controller setpoint. The gain factor for each controller is the ratio of the associated totalized feed setpoint values for the digesters in the respective battery divided by the total feed setpoint values for all batteries (Figure 12.7).

DIGESTER MIXING. Mixing of the contents within the digester is necessary to prevent settling, to develop a homogeneous mass, and to promote heat transfer and contact between the microorganisms and the introduced nutrients.

Mixing can be accomplished by introducing the sludge at different locations around the perimeter of the digester during the feeding cycle, by recir-

---

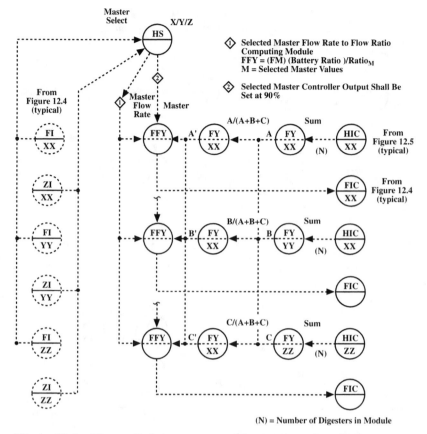

Figure 12.6   Flow splitting process and instrumentation diagram—master/slave

culating the sludge by pumping it from the top of the digester to the bottom (or bottom to top) through the use of mechanical mixers and through the use of draft tubes or *diffusers* using recirculated sludge or digester gas.

Perimeter influent mixing valves may be controlled sequentially using "open" or "close" commands from the logic sequencer. The sequencer is controlled by an operator-adjustable timing function that opens the next valve in sequence before closing the valve currently open to ensure that there is always a path for the feed sludge (Figure 12.8).

Recirculation is performed by pumping sludge from the top of the digester to the bottom or, conversely, from the bottom to the top. This recycled sludge may be introduced into draft tubes within the digester, at multiple injection points around the perimeter, or directly into the digester feed line. The latter provides some additional benefit in that it provides some preheating for the incoming feed. This mixing system can be controlled by a repeat cycle timer but is typically run continuously because the recircula-

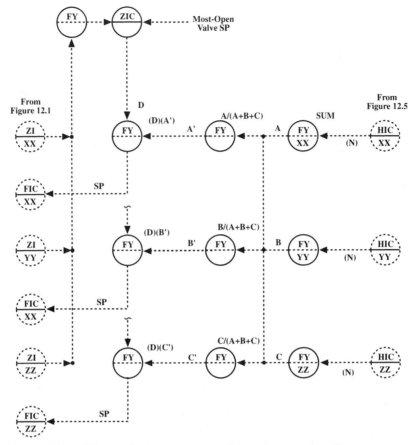

**Figure 12.7 Flow splitting process and instrumentation diagram—
most-open valve**

tion system is often also used to assist with solids withdrawal and/or temperature control (Figure 12.9).

Mechanical mixers use the rotary action of a mixer blade to create an upward or downward flow (depending on mixer rotational direction) of the digester contents. The rotational direction is changed manually or on a timed basis, with the motor controlled by a repeat cycle timer. Rotational changes incorporate a time delay before reversal to prevent damage to the motor. Mixer operation is also interlocked with the digester level to prevent the motor from operating under low-level conditions that might partially expose the blades and cause excessive vibration resulting in mixer bearing or shaft damage (Figure 12.10).

Using the digester gas is another method of providing digester mixing. Gas is withdrawn from the digester dome using a gas compressor and introduced back into the digester near the bottom, either in separate sparger ports or into draft (or "piston") tubes. The injected gas bubbles then rise

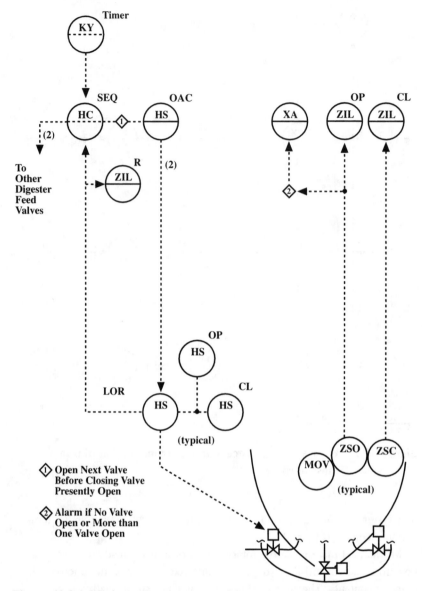

Figure 12.8   Mixing valve process and instrumentation diagram

to the surface of the sludge, thereby generating an upward flow of the digester contents (either unconfined from the sparger or confined through the piston tubes). Gas may be introduced at the center or around the perimeter of the digester, which creates an overall circulation pattern within the digester.

DIGESTER LEVEL MONITORING. With both fixed- and moving-cover digesters, it is important to monitor the level within the digester and,

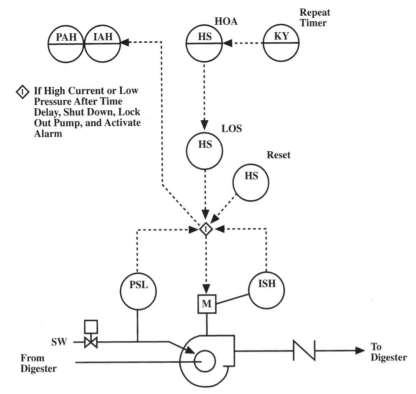

Figure 12.9   Sludge recirculation process and instrumentation diagram

if applicable, the cover. The measurement is useful for determining the solids inventory and various operational parameters used by the operator in evaluating process conditions and regulatory compliance requirements. By using the sludge level and cover positions, an operator can estimate the amount of gas storage within the digester dome. In addition, the level is interlocked with the digester feed and withdrawal routines to prevent over-filling or excessive drawdown. These conditions could cause problems with the mixing apparatus, gas collection and sealing mechanisms, and cover integrity (Figure 12.11).

TEMPERATURE CONTROL. The anaerobic process must be maintained within predetermined temperature limits to ensure efficient digester operation. The temperature is typically monitored by a temperature element inserted through the digester wall or placed in the sludge recirculation line.

Digester heating can be accomplished by introducing low-pressure steam directly into the digester or circulating the sludge through a heat exchanger. With a heat exchanger, heat energy is supplied in the form of hot water or steam and is transferred to the sludge through the walls of the exchanger.

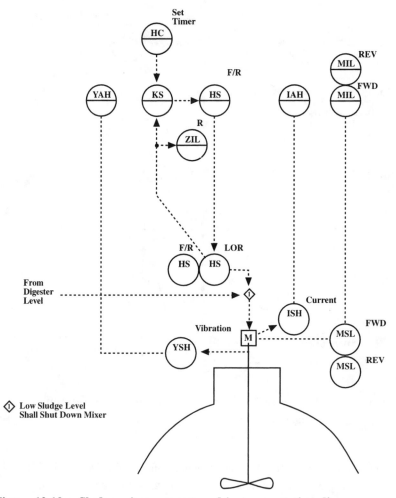

Figure 12.10   Sludge mixer process and instrumentation diagram

This heating method is typically performed by circulating the sludge recycle stream through a heat exchanger with feed sludge added either before or immediately after the heat exchanger. In some cases, the feed sludge is preheated prior to mixing with the heated, recirculating sludge.

To determine the heat energy required to be added to the digester when using steam injection, the flow rate, temperature, and pressure of the steam must be monitored. These data are used along with temperatures of the digester, feed sludge, and ambient air to establish the steam flow rate. The resulting value represents the setpoint for the steam flow rate controller. Controller output modulates the steam control valve to maintain the flow rate established to sustain the digester at the desired temperature.

When using heat exchangers, the flows and inlet and outlet temperatures for both the hot water or steam stream and the sludge stream are moni-

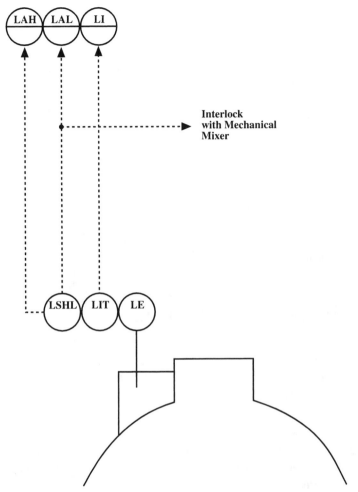

Figure 12.11   Digester level measurement process and instrumentation diagram

tored. Typically, the sludge outlet temperature from the heat exchanger is used as the input signal to a temperature controller, which modulates a bypass valve on the steam or hot water supply line to the exchanger to supply the heat necessary to maintain the digester at the desired temperature. By monitoring the inlet and outlet temperatures and the flows of both heating and heated fluid streams, the heat-transfer efficiencies can be determined. This can be helpful in determining when a heat exchanger is becoming fouled and requires cleaning or other maintenance (Figure 12.12).

DIGESTER SOLIDS WITHDRAWAL. Because of the elevated levels often maintained within anaerobic digesters, withdrawal is typically by gravity. In some digesters, the levels are held constant and the withdrawal is via overflowing a weir; therefore, the volume that goes in equals that which

---

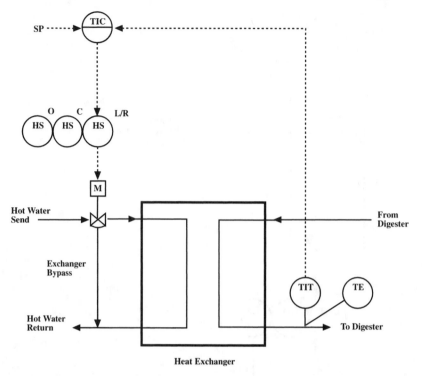

Figure 12.12 Digester temperature control process and instrumentation diagram

comes out. Where withdrawal is not performed by overflow or where pumping is required, the control of the withdrawal process is similar to that of the loading process.

Withdrawal within a battery of digesters is performed sequentially using two-position (opened/closed) control valves on the discharge lines from each digester. Valve (opened/closed) status is provided for operator reference and is used by the sequencer and totalizer to sequence the withdrawal from each digester. The single flow signal is used to totalize the flow from each digester. The flow totalization is performed for each digester when its influent valve is in the open position. The valves are sequenced such that the discharge valve for the next digester is opened as the current digester's discharge valve is being closed. The totalized flow withdrawn from each digester is displayed for the operator's use.

For operations in which the level is permitted to vary, thus providing the operator a limited storage capacity within the digester, the operator must adjust the withdrawal cycle flow setpoint. By using the level measurement and the totalized influent and effluent flows, the operator can determine the required setpoint value to reestablish a desired level within the digester.

DIGESTER GAS MONITORING. The digester gas, a potentially beneficial product of the digester process, is withdrawn from the digester dome. The flow rate is measured, totalized, and recorded. Gas production information may be used as a preliminary indication of potential process problems because the methane-forming bacteria are extremely sensitive to upset conditions. The digester gas header pressure is monitored to provide an indication of problems within the gas collection system, and, if applicable, the gas header pressure is interlocked to shut down the gas mixing system. Overpressurization of a digester can cause structural damage to the cover and/or release of combined sulfide and methane gas to the atmosphere, resulting in odor problems and an explosive atmosphere in the vicinity of the digester.

R*EFERENCES*

Deeny, K.J., *et al.* (1991) Autoheated Thermophilic Aerobic Digestion. *Water Environ. Technol.*, **3**, 65.

U.S. Code of Federal Regulations (1993) Title 40 CFR 503.

U.S. Environmental Protection Agency (1990) *Environmental Regulations And Technology: Autothermal Thermophilic Aerobic Digestion Of Municipal Wastewater Sludge.* EPA-625/10-90-007, U.S. EPA, Washington, D.C.

U.S. Environmental Protection Agency (1992) Environmental Regulations and Technology: Control Of Pathogens and Vector Attraction in Sewage Sludge. EPA-625/R-92-013, U.S. EPA, Office Res. Dev., Washington, D.C.

Chapter 13
Thickening and Dewatering

*G*RAVITY THICKENING

Wastewater sludges are commonly thickened before stabilization and mechanical dewatering to remove excess water and further reduce the volume of the waste. This reduced volume allows the use of smaller, less expensive reactors (or increased capacity of existing basins) and savings in dewatering chemicals. Because gravity thickening is a relatively inexpensive process, the savings resulting from its implementation can be large in comparison to the costs. An additional benefit may be a reduction in variability of the thickened concentration to downstream units, which can result in better control.

In gravity thickening, the particles are thought to actually come into physical contact with each other and form a supporting structure. This process is called *type-4 settling* or *compression settling*. The two factors that primarily determine the final thickened concentration are the residence time in the thickener and the force applied to the particle structure. Increased

---

retention time will increase the solids concentration. Likewise, increased stress resulting from a higher sludge blanket level will increase the final thickened concentration.

PROCESS CONTROL STRATEGIES. Although there is a considerable amount of literature published about the design of gravity thickeners, there is relatively little published about operational control strategies. A brief review of the process shows that the only control parameters available to operators are the influent flow rate and the underflow flow rate. Often, the influent flow rate cannot be controlled, and only the underflow rate can be adjusted. Therefore, it is desirable to implement a control strategy that modulates underflow rate to maximize thickened sludge concentration while minimizing variability of this concentration.

Other objectives of a thickener control strategy may be to minimize clogging of pipes, pumps, and valves; minimize rising solids, which can deteriorate the thickener supernatant; and prevent the generation of odor.

SLUDGE BLANKET LEVEL CONTROL. One simple control strategy uses the sludge blanket level in the gravity thickener as the primary variable for a slow feedback loop. The underflow flow rate is the controller output. Whenever the sludge blanket rises above the level setpoint, the underflow flow rate is increased to remove solids faster and reduce the sludge blanket level. Likewise, when the sludge blanket level is below its setpoint, the underflow flow rate is decreased.

At one particular wastewater treatment plant (WWTP), this strategy was successful in controlling the sludge blanket at 4 m (13 ft), plus or minus 0.3 m (1 ft) (City of Houston WWTP, Texas). Although this strategy was successful at controlling the sludge blanket level, its intended purpose was to control the thickened concentration. It achieved this objective by carrying a high sludge blanket level (4 m [13 ft]) that maximized both the residence time for thickening and the mechanical stress placed on the particle structure. This strategy increased the thickened concentration of waste activated sludge from approximately 3% before implementation to 4% on a consistent basis. Polymer requirements for centrifugal dewatering were greatly reduced.

A process and instrumentation diagram for this control strategy is shown in Figure 13.1. Minimum and optional instrumentation recommendations are listed in Table 13.1. Sludge blanket instruments should generally be mounted approximately halfway between the influent well and the effluent weir to avoid localized disturbances at the influent and the outside wall.

Selection of Sludge Blanket Level Setpoint. The setpoint level for the sludge blanket level strategy must be carefully selected based on a number of site-specific factors. From a thickening point of view, it is desirable to

---

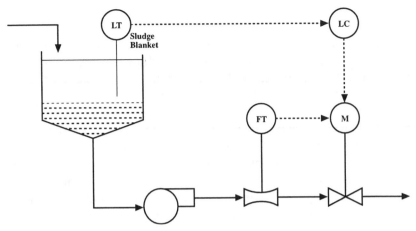

Figure 13.1   Sludge blanket level control of thickening

Table 13.1   Recommended gravity thickening instrumentation

Measurement	Comments
Sludge blanket detector	One needed for each thickener
Flow meter	One needed for each underflow sludge line
Final control element	One needed for each underflow sludge line
Suspended solids concentration	Optional
	Typically a difficult measurement at greater than 2.5% total suspended solids

maintain as high a sludge blanket level as possible. Two potential problems, however, may limit the amount of sludge that can be held. An appropriate sludge blanket level setpoint is typically determined experimentally based on the following factors.

The sludge blanket level must be kept sufficiently low that the thickener can accommodate normal fluctuations in the influent flow. If the sludge blanket level is too high, a sudden increase in flow rate, causing the sludge blanket to rise, may cause solids to be swept out in the thickener supernatant, resulting in a failed thickener.

If solids are held too long in the thickener, gas bubbles may be generated that rise to the surface. These bubbles often carry large amounts of solids into the supernatant and disturb the thickening process in general. The gas may be the result of anaerobic decomposition or, more commonly, of denitrification. Denitrification occurs when facultative bacteria are presented with a readily biodegradable food source, a lack of dissolved oxygen, and a nitrified effluent with nitrate (NO_3^-). Under these conditions, the bacteria use the chemically bound oxygen and release nitrogen gas. These conditions can be easily realized if primary sludge is cothickened with a nitrifying waste activated sludge.

Sludge Blanket Detectors. There are currently two types of instruments for measuring sludge blanket level. One type uses an optical sensor to measure suspended solids. The sensor probe itself is suspended on a cable rolled up on a drum. The sensor is lowered by rotating the drum and unfurling the cable. The sludge blanket level is calculated by measuring the number of turns of the drum before the optical instrument measures a suspended solids concentration greater than a user-determined level. Some units provide a continuous reading of the blanket level, while others provide periodic readings. Some units can also be programmed to perform a solids profile of the entire thickener.

Optical sludge blanket detectors may have high maintenance requirements because they contain both moving parts and electronic components. The connection between the optical probe and the electronic unit has historically been a weakness of this instrument. Sliding connections tend to corrode over time, while hard-wired connections tend to twist and break. The other basic components of the instrument are reliable.

Optical sludge blanket detectors must also contend with a rotating collector mechanism that can snag the suspended probe. A contact switch is required either to raise the probe out of the water when the collector comes around or to initiate a cycle after the collector passes.

In recent years, ultrasonic sludge blanket instruments have been introduced. In principle, they are similar to the fish finders available for fishermen. They measure the sludge blanket interface by bouncing sound waves off of it. If the application has a clearly defined interface with a significant difference in density between the two layers, these instruments can work well. Biological solids, however, often have a density similar to water with poorly defined blanket interfaces. Experimentation at each specific site is recommended.

Ultrasonic sludge blanket instruments have the potential for low installation and maintenance costs. The ultrasonic probe is mounted just under the water surface on a hinged length of conduit. It simply swings on the hinge over the scum and collector mechanisms.

SLUDGE CONCENTRATION CONTROL. Another simple control strategy uses the thickened solids concentration as the primary variable for a slow feedback loop, as shown in Figure 13.2. The underflow flow rate is the controller output. Whenever the concentration rises above the setpoint, the underflow flow rate is increased to remove solids faster and reduce the

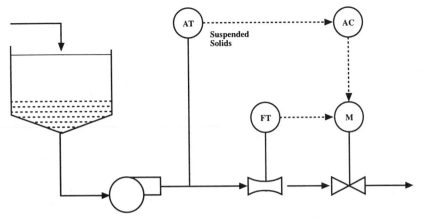

Figure 13.2   Sludge concentration control of thickening

concentration. Likewise, when the concentration is below its setpoint, the underflow flow rate is decreased.

While this control strategy is simple and has the potential to produce a high concentration with little variability, the solids concentration measurement is the major perceived weakness. While this measurement does not have to be accurate, it must be at least repeatable. Recent tests by the non-profit Instrumentation Testing Association (1993) showed that many, but not all, of the commercially available instruments fall short of these expectations. Major problems include the large amount of oil and grease in primary sludge that fouls optical instruments, stringy materials that bind mechanical elements, and entrained gas that distorts optical properties and density. Experimentation at each specific site is recommended.

CENTRIFUGES

Centrifuges are used to either thicken or dewater at WWTPs. Centrifuges use centrifugal force to separate solids from a solid–liquid slurry. Solid particles with densities greater than the suspending fluid move away from the axis of rotation, causing a separation between the dense solid particles and the lower-density particles and liquids (*centrate*).

The centrifuge process equipment systems used for thickening or dewatering are similar. Figure 13.3 shows the process flow diagram for a typical thickening or dewatering countercurrent solid bowl centrifuge. Sludge mixed with polymer is fed to the centrifuge. The centrifuge separates the solids from the liquids. The centrate is recycled back to the WWTP for processing. Typically, a conveyor or pump is used to transfer thickened or dewatered solids to a storage tank or bin. A lubrication system is generally supplied with the centrifuge to lubricate the main bearings supporting the

---

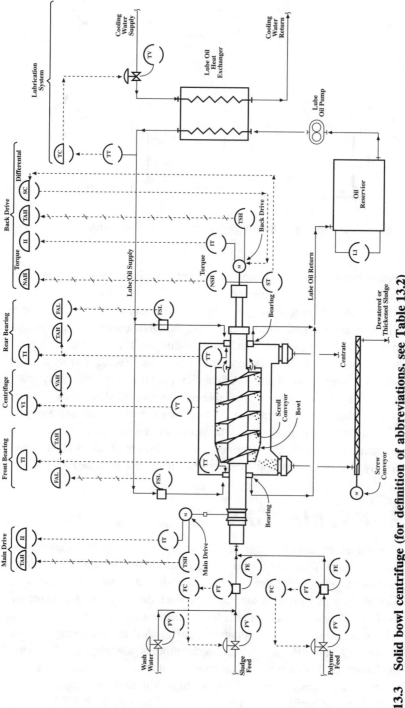

Figure 13.3   Solid bowl centrifuge (for definition of abbreviations, see Table 13.2)

centrifuge. The lubrication system typically consists of an oil reservoir, an oil pump, and a heat exchanger.

Control strategies discussed in this section will be oriented toward solid bowl centrifuges because they are most frequently used in WWTPs. A more complete description of the operation of solid bowl centrifuges for thickening and dewatering is given in the Water Environment Federation's Manual of Practice No. 11, *Operation of Municipal Wastewater Treatment Plants* (WEF, 1996). Controls and instrumentation shown in Figure 13.3 will be discussed throughout this section.

Control strategies used for centrifuge thickening or dewatering should have the following goals:

- Centrifuge performance:
  — Maximize thickened or dewatered solids recovery,
  — Minimize solids concentration in centrate, and
  — Minimize polymer usage.
- Machine control:
  — Minimize down time (preventive maintenance), and
  — Facilitate personnel and equipment safety.

The primary variables that affect the centrifuge performance are solids characteristics and feed rate, polymer feed, and differential speed between the centrifuge bowl and the scroll conveyor. Solids characteristics, including particle size and concentration, are generally controlled by upstream processes, clarifiers, and digesters. Fine particles will require more polymer to agglomerate. Large or dense particles are more easily separated by centrifugal force. Feed rate, both mass and volume, must be controlled to match both the hydraulic-loading and the solids-conveying capacity of the centrifuge. Solids concentration variations affect loading, which is limited by the maximum conveying capacity of the centrifuge. The differential speed between the bowl and the scroll controls the conveying capacity of the thickened or dewatered solids and also affects both capture and cake concentration.

Control strategies discussed in this section are

- Feed flow control,
- Polymer flow control, and
- Machine control.

CONTROLLING FEED FLOW. Feed is controlled to match the hydraulic- and solids-conveying capacities of the centrifuge. Feed flow can be controlled manually or automatically. For manual feed control, the flow rate is controlled by adjusting a valve or pump speed. Hydraulic variations and/or sludge characteristics changes will cause the flow to vary over time from the manual setpoint. Automatic control of feed is shown in Figure 13.3. A controller compares the feed rate to a setpoint and modulates a control element to control the flow. Table 13.2 lists the minimum instrumentation requirements for controlling feed. Accurate feed control will improve centrifuge performance.

CONTROLLING POLYMER FLOW. Polymer is used to condition the sludge so that particles agglomerate and form larger or denser particles that can be separated easily by the centrifuge. Polymer usage is a function of the feed characteristics. Variations in the ratio of primary to secondary sludge, temperature, and particle size affect the polymer usage. An objective of a good control strategy for polymer control is to minimize polymer usage while maximizing the solids retention of the centrifuge-dewatered or centrifuge-thickened sludge.

A simple control strategy for polymer addition is flow control as shown in Figure 13.3. Polymer flow control has been successfully used for many centrifuges. Automatic polymer flow control works similarly to feed control as described previously. The setpoint that is chosen is typically based on operation experience and visual inspection of the dewatered or thickened sludge and centrate. Table 13.2 lists the minimum instrumentation requirements for polymer flow control.

Advanced methods for controlling polymer flow measure the properties of the conditioned solids or centrate. These measurements can be used as inputs to a cascade controller that adjusts the setpoint of the flow control loop. One such method uses a streaming current detector that measures the electrical charge of particles in the centrate. The electrical charge in the centrate is proportional to the number of particles. An increase in the measured charge corresponds to a poor solids capture, requiring an increase in polymer dosage to flocculate the particles. This strategy is used to minimize the amount of particles in the centrate or to maximize solids capture by adjusting the polymer flow.

Another method of measuring centrate uses a turbidity analyzer. An increase in suspended solids in the centrate causes an increase in the turbidity of the centrate. This strategy is also used to minimize the amount of particles in the centrate or to maximize solids capture by adjusting the polymer flow. Foaming and entrained air in the centrate are obstacles to implementing suspended solids measurement.

MACHINE CONTROL. Centrifuge machine control strategies are designed to optimize performance, to properly start up and shut down the equipment, and to monitor operation of centrifuge and support equipment for maintenance and safety. Machine control varies widely among manufacturers of centrifuges. The control strategy descriptions in this section

Table 13.2 Minimum recommended instrumentation for centrifuge control (abbreviations used in Figure 13.3 are shown in parentheses)

Measurement	Comments
Sludge flow control (FC)	
Flow rate (FT) (FE)	Sludge feed flow measurement
Control element valve (FV) or variable-speed pump	Sludge feed flow control
Polymer flow control (FC)	
Flow rate (FT) (FE)	Polymer feed flow measurement
Control element valve (FV) or variable-speed pump	Polymer feed flow control
Start-up and shutdown sequences	
Hand switches	For starting each drive, pumps, and conveyors, and automatic start-up and shutdown sequence
Main drive status	On–off indication from motor starter auxiliary contact or current switch
Back drive status	On–off indication from motor starter auxiliary contact or current switch
Lube oil pump status	On–off indication from motor starter auxiliary contact or current switch
Screw conveyor status	On–off indication from motor starter auxiliary contact or current switch
Sludge feed pump status	On–off indication from motor starter auxiliary contact or current switch
Polymer feed pump status	On–off indication from motor starter auxiliary contact or current switch
Back drive speed indication	Scroll shaft speed indication
Wash water valve	For washing the centrifuge on shutdown
Wash water valve position	Open and closed valve position switches

continued

Table 13.2   Minimum recommended instrumentation for centrifuge control (abbreviations used in Figure 13.3 are shown in parentheses) (continued)

Measurement	Comments
Machine protection and maintenance monitoring	
Centrifuge vibration transmitter (VT)	Vibration monitoring and high vibration for monitoring (VI), alarms (VAH), and centrifuge shutdown
Main bearing temperatures transmitters (TT)	Temperature should be monitored on all main bearings with high temperature alarms (TAH) and centrifuge shutdown setpoints
Lube oil low-flow switches (FSL)	Low lube oil flow to each main bearing for alarming (FAL) and centrifuge shutdown
Motor winding temperature switches (TSH)	On both the main and back drive motor windings for alarm (TAH) and centrifuge shutdown
Motor current transmitters (IT)	On both the main and back drive motor windings for current (II) or power indication and recording; back drive current signal is also used for automatic torque control
High back drive torque switch (NSH)	Alarm and shutdown centrifuge on high torque (NAH)
Lube oil temperature transmitter (TT)	For temperature control of lube oil
Lube oil cooling water control valve (TV)	Control cooling water flow to heat exchanger
Lube oil reservoir oil level indicator (LI)	Monitor oil level for lubrication system
Emergency stop switches and pull cords	Locate around equipment for personnel safety

should be used as guidelines for designing and selecting features when integrating centrifuge control into WWTP operations.

Differential Speed Control. Differential speed between the bowl and the scroll conveys the solids from the feed zone to the discharge. The differential speed is set to match the solids loading in the feed. Differential speed has been controlled by several methods, including gear boxes with dual gear reductions for the scroll and bowl or eddy current brakes on the scroll. Eddy current brakes are used for automatic control of differential speed. The differential speed can be controlled automatically in two modes: constant and variable differential speed. Constant differential speed is typically controlled by measuring the scroll shaft speed and adjusting the drive current using a feedback controller. Constant-speed control works well for thickening and dewatering centrifuges that have constant sludge characteristics.

Variable-speed control is also known as *autotorque* control. In autotorque control, the differential speed is varied between preset limits to maintain a constant torque on the scroll. Torque is exerted on the scroll by the friction caused by conveying the solids to the discharge chute. Current to the scroll drive is proportional to the torque exerted on the scroll drive shaft. The scroll drive current is measured and controlled to maintain a constant torque. As the sludge characteristics vary, the scroll speed varies, maintaining a constant torque. Automatic torque control works well for dewatering centrifuges, providing a dewatered cake with maximum solids concentration.

Start-Up Sequence. Manual and/or automatic start-up sequences should be part of the control strategies developed for the centrifuge. A typical start-up sequence for the centrifuge shown in Figure 13.3 is as follows:

1. Start lube oil pump.
2. Start sludge discharge screw conveyor.
3. When the lube oil pump and the discharge screw are confirmed running, start the main drive and start the back drive at minimum speed.
4. After the centrifuge is at its operating speed, enable polymer and feed to the centrifuge. This may require starting sludge and polymer feed pumps.

After the system is started, the feed, polymer, and scroll speed setpoints should be adjusted to obtain the maximum thickened or dewatered solids concentration. Status indication of lube oil pump, screw conveyor, main centrifuge drive, centrifuge back drive, and sludge and polymer feed pumps will be required as a minimum for the start-up sequence. Centrifuge back drive speed indication is also required.

Shutdown Sequence. As with the start-up sequence, a manual and/or automatic shutdown sequence is required. A typical shutdown sequence for the centrifuge shown in Figure 13.3 is as follows:

1. Stop the polymer and feed to the centrifuge.
2. Set the scroll drive to minimum speed.
3. Open the wash water valve and wash the centrifuge for a set period of time.
4. Stop centrifuge bowl and scroll drives.
5. Stop screw conveyor.
6. When the centrifuge has stopped moving, stop the lube oil pump.
7. Rinse centrifuge for a set period of time and then close the wash water valve.

The minimum instrumentation is listed in Table 13.2 for start-up and shutdown sequences.

Machine Protection and Maintenance Monitoring. Machine protection and maintenance monitoring should be part of the centrifuge control strategy to protect the investment in the equipment and to provide safe operation. The following signals should be interlocked to shut down the centrifuge for machine protection:

*   Lube oil pump failure,
*   High centrifuge vibration,
*   High temperature for main bearings,
*   Low lube oil flow to main bearings,
*   Main and back drive high winding temperatures,
*   High torque on scroll,
*   Discharge screw conveyor or pump failure, and
*   Emergency stop pull cord or switch.

Maintenance monitoring should be done to monitor changes over time in the machine performance. Appropriate instrumentation for machine protection and maintenance monitoring is listed in Table 13.2. Changes to the variables monitored indicate wear on the bearings and scroll blade. Monitoring and recording these variables will provide an indication of the need for preventive maintenance, which will increase the machine life and prevent unscheduled downtime. As a minimum, the following variables should be monitored and recorded:

*   Centrifuge vibration level,
*   Bearing temperatures, and
*   Motor current or power consumption.

*B*ELT PRESS FILTRATION

Belt presses continue to represent a significant method of dewatering in the municipal wastewater sector. Although the industry is looking to other alternatives when higher cake solids are required, belt presses continue to be popular because of their simplicity of operation and their relatively low capital cost. They are offered in a wide variety of configurations, but all state-of-the-art models have three common elements or processing zones.

In the gravity dewatering zone, flocculated sludge is deposited on the surface of a belt, where free water drains through the belt under the force of gravity. "Ploughs" are typically employed to turn the sludge in a manner that continuously exposes fresh drainage surfaces to the belt and thus enhances the drainage rate of the water.

In the second zone, the sludge is exposed to low or medium pressure as the cake enters a wedge that is formed between two converging belts. The pressure is applied on both sides of the cake, and water drains out through both belts.

The final zone is one of high pressure, where the two belts containing the cake are guided around a series of various-sized rollers. This serpentine arrangement not only increases the direct pressure on the sludge but also subjects the cake to shear forces because of the differences in the radii of the two belts passing around the rollers. At the end of the high-pressure zone, the cake is discharged by means of doctor blades or scrapers mounted at the surface of both belts, and the belts are washed before reentering the gravity zone.

The characteristics of municipal sludge are such that the need for chemical conditioning, typically with organic polyelectrolytes, to enhance the rate of water removal is essentially a foregone conclusion. This is particularly true when biological solids represent a significant fraction of the total. As with other dewatering devices, polymer addition is probably the single most important factor in determining the performance of a belt press. The polymer requirement at any instant in time is based on the mass flow of incoming solids and some gross parameter generally defined as the *dewaterability* of these solids. The solids mass flow will vary if the liquid flow rate or the solids concentration of the feed vary either independently or simultaneously. The dewatering characteristics, or dewaterability, will also vary over time but not necessarily in either a predictable or measurable fashion. A primary objective of any automatic polymer control system should therefore be the ability to identify and to respond to variations in the liquid flow rate, the solids concentration, and the dewaterability.

Although it is generally accepted by WWTP operators that there is a need to automatically control polymer addition to belt presses, it is difficult to find documented examples of full-scale operating systems. Proprietary

---

polymer control systems are beginning to be offered by belt press manufacturers as integral components of their presses. For example, one manufacturer offers a system that uses optical sensors to measure the light reflectance from the surface of the sludge in the gravity dewatering zone. In their brochure, they indicate polymer savings of 30% over manual operation. Other companies are also addressing the issue of polymer control, but application information, other than promotional literature, is difficult to find.

The simplest form of polymer control is to supply the polymer at a flow rate that is proportional to the incoming feed rate. This allows the operator to determine experimentally the ratio that will minimize polymer consumption without adversely affecting dewatering performance. However, the success of this approach is dependent on the assumption that a homogeneous sludge, in terms of solids concentration and dewatering characteristics, can be maintained for extended periods of time. Unfortunately, the inability to maintain a homogeneous sludge is the primary reason that a more sophisticated control approach is required. Dewatering characteristics and solids concentrations do change for a variety of reasons and significantly affect dewatering performance and polymer consumption.

One strategy for the automatic control of polymer addition to a belt press, which has been successfully demonstrated and is well documented in the literature, involves on-line measurement of the rheological characteristics (that is, the properties of matter in the fluid state) of the flocculated sludge just before its discharge to the gravity dewatering zone (Campbell and Crescuolo, 1984 and 1989; Campbell *et al.*, 1986; and Crawford, 1990). The basic control loop is shown in Figure 13.4 (Crawford, 1990). The control strategy operates by first "tuning" the press. In this phase, the operator manually adjusts the polymer dosage to the minimum level that results in acceptable performance of the press, that is, with no sludge squeezing out between the belts in the wedge zone. Five consecutive samples are taken, and the measured rheological characteristics are logged and averaged by the microprocessor to establish the setpoints for the system. In the automatic control mode, the system takes samples according to a variable predetermined frequency and measures the rheological characteristics. These characteristics are compared to the setpoints stored in the computer, and if they are different, the control algorithm initiates a change in the polymer flow rate. A minimum-seeking algorithm is used to ensure that the system continually moves toward the minimum level of polymer that will satisfy the setpoints in the memory.

An important feature of this approach is that, within certain boundary conditions, the system does not differentiate between effects caused by changes in the mass flow of solids and the dewatering characteristics of the sludge, but simply recognizes the overall effect on the quality of the flocculated sludge and initiates the appropriate control action. Figure 13.5 (Campbell and Crescuolo, 1989) illustrates the results of the system in operation

---

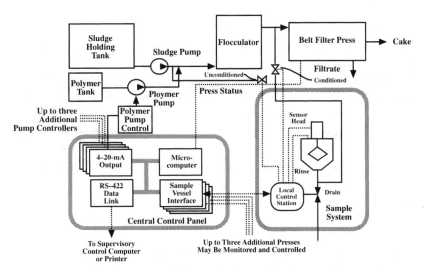

Figure 13.4   Automatic polymer addition control loop (reprinted from
Crawford, P.M. (1990) Optimizing Polymer Consumption
in Sludge Dewatering Applications. *Water Sci. Technol.*,
22, 261, with permission from Elsevier Science Ltd, The
Boulevard, Langford Lane, Kidlington OX5 1GB, UK).

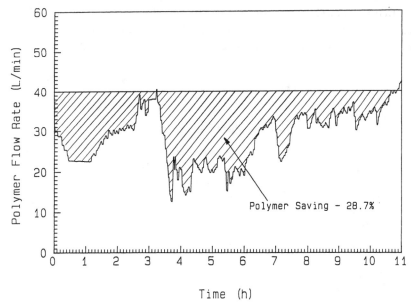

Figure 13.5   Response of automatic polymer control system to variable
feed solids (reprinted from Campbell, H.W., and Cres-
cuolo, P.J. (1989) Control of Polymer Addition for Sludge
Conditioning: A Demonstration Study. *Water Sci. Technol.*,
21, 1309, with permission from Elsevier Science Ltd, The
Boulevard, Langford Lane, Kidlington OX5 1GB, UK).

---

on a two-meter belt press. The horizontal line represents historical manual operation where the polymer flow rate was essentially set in the morning and left unchanged for the rest of the day. The fluctuating line represents the polymer flow rate as the control system tracked the changing characteristics of the incoming sludge. In this instance, the polymer savings between historical manual control and automatic control was approximately 29%.

The use of a streaming current detector has also been demonstrated as a viable conditioning control strategy for belt presses (Dentel and Abu-Orf, 1994). Streaming current is a measure of the *zeta potential*, or electrophoretic mobility of the colloidal material in the filtrate, and a number of researchers, summarized by Dentel and Abu-Orf (1994), have shown that a near-zero streaming current corresponds to good dewaterability. Dentel and Abu-Orf evaluated the potential for using streaming current for automatic control of the polymer feed to a full-scale belt press. A schematic drawing of the sampling apparatus is shown in Figure 13.6. The main conclusion of the study was that a near-zero filtrate streaming current indicated good dewaterability and was successfully used as the streaming current detector's setpoint for automatic control purposes. Thorough cleaning of the apparatus was required at least once a week to avoid a gradual drift of the device's output. No estimates of the projected polymer savings resulting from automatic control were reported.

Belt press machine controls that can be varied include belt speed, belt tension, and the degree of mixing in some kind of flocculation system, either a stirred vessel or an in-line system. Flocculation systems are gener-

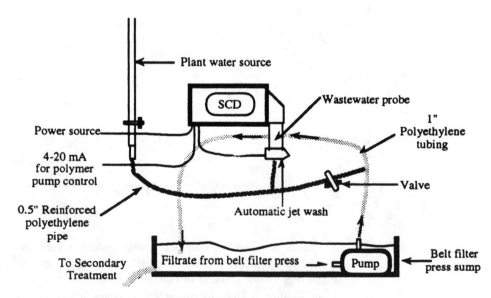

Figure 13.6   Schematic diagram for streaming current detector (SCD) filtrate sampling (Dentel and Abu-Orf, 1994)

ally specific to a particular belt press manufacturer and not integrated into the automated control strategy. The optimum belt tension is generally determined based on considerations of the type of sludge, the level of cake solids desired, and the acceptable operating life of the belt. The belt tension is not amenable to on-line adjustment. Conversely, the belt speed is easily varied on line and in Europe has been incorporated into a control strategy (Pieters, 1992, personal communication, Zhen Environmental b.v., Purmerend, The Netherlands). An ultrasonic sensor is used to monitor the level of the sludge in the gravity dewatering zone. A decrease in the level, resulting, for example, from a decrease in feed solids, would result in a decrease in the belt speed such that the mass of solids to the pressure zone would remain constant. It is suggested that this would result in more consistent cake solids and also maximize the throughput capacity of the press. No performance results are available in the literature for this technology.

Another European company is offering a package that is advertised as an automatic belt press management and control system. The basis of the control strategy is the measurement of filtrate turbidity by light absorption. This package provides data logging and some data manipulation and can control polymer and sludge flow rates. Although no performance data are presented, it is claimed that polymer reductions of up to 40% have been achieved.

The benefits of automated control for belt presses, at the current stage of development, are essentially related to polymer control. An effective polymer control strategy eliminates the need for overdosing and thus reduces polymer costs. At the same time, the potential for press failure as a result of underdosing is also reduced; therefore, the need for operator attention is minimized, resulting in savings in operation labor. Additional benefits, which are generally accepted as being possible but have not been fully documented in the field, include less variation in the cake solids and increased press capacity.

Although polymer selection is not, strictly speaking, a belt press control parameter, it is obviously a critical element in the development of efficient and cost-effective performance. Selection of the most effective polymer can be an extremely complicated and frustrating procedure. A recent publication (Dentel et al., 1993) provides a practical step-by-step approach to both selecting the most appropriate polymer and evaluating its performance in full-scale application. This manual represents an excellent first step in developing a strategy to address a problem that is often considered to be more of an art than a science.

REFERENCES

Instrumentation Testing Association (1993) *Performance Evaluation of High Range Suspended Solids Analyzers For Wastewater Treatment Applications.* Washington, D.C.

Campbell, H.W., and Crescuolo, P.J. (1984) Development of a Sludge Conditioning Control Strategy Based on Rheological Measurements. Paper presented at the 57th Annu. Conf. Water Pollut. Control Fed., New Orleans, La.

Campbell, H.W., and Crescuolo, P.J. (1989) Control of Polymer Addition for Sludge Conditioning: A Demonstration Study. *Water Sci. Technol.,* **21**, 1309.

Campbell, H.W., *et al.* (1986) An Instrument for Automated Control of Sludge Conditioning. Paper presented at the 9th Symp. Wastewater Treat., Montreal, Que., Can.

Crawford, P.M. (1990) Optimizing Polymer Consumption in Sludge Dewatering Applications. *Water Sci. Technol.,* **22**, 261.

Dentel, S.K., *et al.* (1993) *Guidance Manual for Polymer Selection in Wastewater Treatment Plants.* Water Environ. Res. Foundation Project 91-1SP-5, Alexandria, Va.

Dentel, S.K., and Abu-Orf, M.M. (1994) Full-Scale Evaluation of Available Sludge Conditioning Control Technologies. Paper presented at the 67th Annu. Conf. Water Envir. Fed., Chicago, Ill.

Water Environment Federation (1996) *Operation of Municipal Wastewater Treatment Plants. Volume III: Solids Processes.* Manual of Practice No. 11, Alexandria, Va.

Chapter 14
Disinfection

Disinfection of effluent before discharge is generally required to meet wastewater treatment plant (WWTP) permit limits. The purpose of the disinfection is to ensure protection of public health by eliminating pathogenic organisms that may have survived through the treatment process. Because it is impossible to test for all potential pathogens, the general practice is to use fecal coliform as an indicator organism for judging the effectiveness of the disinfection process. This choice is based on the fact that fecal coliform bacteria are present in wastewater in high numbers and that they tend to be hardy organisms. Consequently, it is presumed that a low fecal coliform count is evidence of the reduction of pathogens to a safe level from a public health standpoint.

A number of processes can be used for disinfection. The most common of these is the addition of a strong oxidant such as chlorine. Because chlorine is toxic to many aquatic organisms, alternative disinfectants such as chlorine dioxide and bromine dioxide have been used in some cases. Ozone has also been used to achieve disinfection. When chlorine is used, it is frequently necessary to remove excess chlorine through the use of a dechlorination process, once acceptable levels of pathogen reduction have been achieved. In recent years, ultraviolet (UV) lamps have been used as alternatives to the addition of oxidants to achieve disinfection. Control strategies for these processes are discussed below.

CHLORINATION CONTROL

There are a number of strategies that can be used to control chlorination. The choice of the strategy to be used is dependent on a number of factors. The primary physical considerations are the means of chlorine feed and its responsiveness, the mixing regime used to distribute the chlorine in the liquid stream flow, and the location of the sampling site with respect to the injection site. In addition, the wastewater characteristics with respect to reactivity to chlorine may also play a role in dictating the strategy used. It should be kept in mind that the goal of chlorination control is to achieve acceptable levels of effluent disinfection, which in most cases is determined by the final fecal coliform levels. As a consequence, the fecal coliform count, which is not available until 1 to 2 days after sampling, is the primary control parameter, with chlorination control setpoints being adjusted on a real-time basis to achieve on-line results. Operating experience determines the chlorine residual at which the fecal coliform counts are within the permitted levels. Because this concentration will vary over time as the flow, temperature, and composition of the effluent change, it is necessary to adjust the chlorine residual setpoint based on the results of fecal coliform testing.

Before discussing the control strategies that are available, it is appropriate to discuss the factors that help dictate the type of strategy used. The first item to be considered is the type of system used to deliver chlorine. Some systems, such as the direct feed of a hypochlorite solution or the use of a gas system to deliver chlorine to the application site under vacuum, provide rapid response times following a change in the feed signal. Other systems, such as ones that produce a chlorine solution and then pump that solution some distance to the application site, have a much slower response time to changes in the feed control signal.

Chlorine mixing efficiency also plays a significant role. The ideal situation is a high-energy mixing regime that approaches instantaneous mix conditions. This is advantageous both from the standpoint of system control and also from the standpoint of chlorination efficiency. In general, the poorer the mixing conditions, the less efficient the chlorination process and the poorer the response time of the system will be. Poor mixing can result in localized areas of high chlorine concentration, which can result in an increased chlorine demand and reduced chlorination efficiency because of undesired side reactions. These can include oxidation and/or chlorination of organic compounds in the effluent and conversion of ammonia to dichloramine, trichloramine, or nitrogen gas resulting from breakpoint chlorination reactions.

The location of the sampling point with respect to the chlorination point has a direct effect on the response time of the system. Location of the sample site too far downstream of the chlorine feedpoint can result in an excessive lag time between a control adjustment and the measurement of its effect. This will require the use of long-time constants in the control strategy and will cause a tendency for the residual to wander. Conversely, locating the sample site too close to the application point can result in collection of a sample in which the immediate uptake has not fully occurred, thus leading to erroneous sample results and unstable control. As a rule, sampling of the effluent for chlorine residual analysis should take place at a point 1 to 3 minutes downstream of the injection point. This assumes that injection is followed by rapid mixing.

Finally, the chemical characteristics of the process stream being treated can have a significant impact because of the effect that this can have on the chlorine uptake rate. Typical domestic wastewater has a chlorine uptake rate that is initially rapid and tapers off quickly. This means that the sampling point can be located close to the injection point without compromising the sample analysis. In some cases, especially with industrial effluents, there may be compounds present that have relatively slow reaction rates with chlorine. This can have an effect on the best sampling point location and could require that sampling be located at some distance removed from the chlorine dosing point. The pH of the wastewater can also affect the reaction rate. Chlorine dissolved in water hydrolyzes to form hypochlorous acid. Hypochlorous acid can then dissociate to form hypochlorite (OCl^-) and hydrogen ion. The proportion of hypochlorous acid to hypochlorite is a function of pH, with a low pH leading to a higher proportion of hypochlorous acid, which is the more reactive form. Consequently, increasing the pH can reduce the chlorination efficiency.

Another factor that affects the control strategy selected is the parameter used to monitor the chlorination process. The most common parameter is chlorine residual. This can be measured either manually or automatically using on-line instrumentation. An oxidation-reduction potential (ORP) procedure can be used as an alternative control parameter to chlorine residual, provided that measurement of chlorine residual is not required by the WWTP permit. The most important consideration is that the selected control parameter correlate with permit requirements. As discussed previously, this is often a fecal coliform limit; however, many WWTPs are required to monitor and/or control chlorine residual as well.

A number of potential strategies are available for the control of chlorination. These range from the simplest manual measurement and adjustment approaches to compound-loop control systems. The specific strategy implemented will depend on a number of considerations. Generally, the complexity and amount of automation will increase as the size of the WWTP or the sensitivity of the receiving water increases. Below is a review of several of the most common control strategies, along with a discussion of their strengths and weaknesses. In all cases, the general approach is to establish and maintain a control setpoint that results in a final fecal coliform count within the desired range. Typically, this range is established by the WWTP's National Pollutant Discharge Elimination System permit.

All of the strategies involve the modulation of the chlorine feed rate, either manually or automatically. The method used to control the chlorine feed depends on the chlorine system being used. If a hypochlorite solution is being used, feed control can be accomplished by means of a metering pump or other volumetric metering system. If liquid chlorine is being used, the basis of the metering system is a standard chlorinator that is capable of modulating the chlorine flow based on an analog input signal and of concurrently generating an analog signal proportional to the chlorine feed rate. The most responsive system involves the transport of chlorine gas to the application site under vacuum using an eductor, with the resultant chlorine solution being injected into the wastewater. An alternative approach involves the generation of a chlorine solution adjacent to the chlorinator, again using an eductor, with the resultant chlorine solution being piped to the application site. A disadvantage to this latter approach is that there is a significant delay between a change in the chlorine feed rate and the time when the altered concentration solution arrives at the injection point. This has a significant effect on the response of the control system chosen and the tuning of that system.

MANUAL CHLORINE CONTROL. The simplest control strategy involves periodic manual testing of the effluent to determine its chlorine residual and manual adjustment of the chlorine feed rate to achieve the desired residual. This approach requires a minimum of instrumentation, but it is only practical in small WWTPs. It does not provide any means of compensating for changes in WWTP flow rate or in the chlorine demand of the effluent. As a consequence, it typically results in a significant level of chlorine overdose to ensure compliance with the WWTP coliform limits. This system is illustrated in Figure 14.1.

MANUAL MEASUREMENT WITH FLOW PROPORTIONING. A variant of the complete manual control scheme described above involves the use of a flow ratio control. With this system, the chlorine feed rate is paced according to the flow rate, with periodic measurement of the chlorine residual being used to adjust the ratio setpoint to maintain the desired effluent residual level. This approach provides better control because it compensates for changes in the WWTP flow rate, but this method still fails to respond to changes in the chlorine uptake rate during the times between manual sampling of the residual. As with the previous strategy, this approach generally requires that a significant level of chlorine overdosing

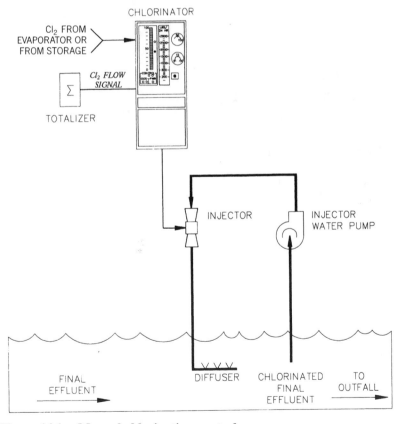

Figure 14.1   Manual chlorination control

be maintained to ensure compliance with permit limits. An example of this system is shown in Figure 14.2.

METHODS INCORPORATING ON-LINE MONITORING. A number of control strategies incorporate the on-line measurement of chlorine residual. There are several chlorine residual analyzers on the market. In general, they all require a reasonable investment of time for maintenance and calibration to ensure their reliable operation. It is important to the success of any of these strategies that proper maintenance, cleaning, and calibration of the analyzers and other equipment be provided. As indicated earlier, alternative process variables such as ORP may also be used as a part of these strategies. Chlorine analyzers for this application have been tested by the Water and Wastewater Instrumentation Testing Association of North America (1990).

STRAIGHT RESIDUAL CONTROL. The use of chlorine residual to provide direct control of the chlorine feed rate is the simplest control strategy of this type. This strategy has the ability to respond to changes both in flow

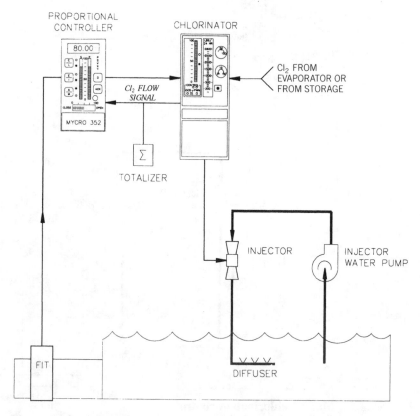

Figure 14.2   Manual measurement with flow proportioning

and in chlorine demand in the wastewater. For this strategy to be successful, a number of conditions must be met. Mixing of the chlorine with the wastewater must occur as close to instantaneously as possible, and it must be possible to obtain a sample within 3 to 5 minutes downstream of the injection point. In addition, the chlorine injection method must be capable of providing close to instantaneous response to changes in the feed control signal. Where these conditions can be met, this control strategy can be reliable and successful. As discussed previously, the success of this strategy requires the routine analysis of the effluent for fecal coliform (or whatever other indicator is identified in the permit), with the chlorine control setpoint being determined by correlating the bacterial count with the measured chlorine residual. An example of this control scheme is presented in Figure 14.3.

COMPOUND CONTROL LOOPS. When conditions of rapid mixing, rapid response to chlorine feed signal changes, and sampling immediately downstream of the injection point cannot be achieved, compound control loops that incorporate both flow and chlorine residual measurement are often used. There are several versions of this type of strategy. One type of

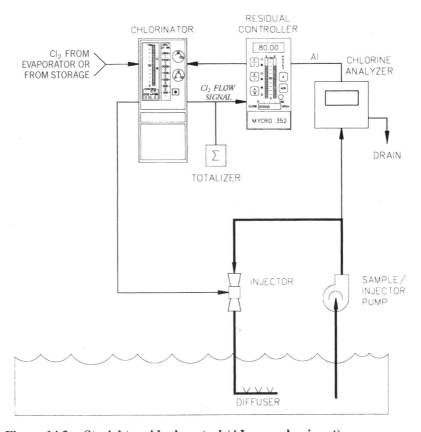

Figure 14.3   Straight residual control (AI = analog input)

---

chlorinator allows both signals to be used to control chlorinator valve position and vacuum. Another variant combines the two signals in a multiplier and uses the output of the multiplier as the input signal to the chlorine feed control device. Another option uses a cascade control loop, with the flow signal being used to generate a control signal with a short reset time. The control signal is fed to a ratio controller that is adjusted by the chlorine residual signal using a long reset interval. This approach has the advantage of allowing immediate response to flow changes while allowing chlorine residual to be used to trim the feed rate when system response times do not allow direct residual control. A control loop that is dependent on flow is subject to failure in the case of a loss of the flow signal. This can be overcome by supplying an average control signal to the controller in the event that a loss of the flow signal is detected. The control strategy then reverts to a straight residual control mode as described above. An example of this compound control loop system is presented in Figure 14.4.

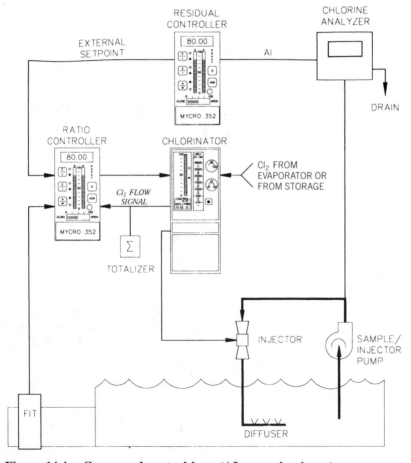

Figure 14.4   Compound control loop (AI = analog input)

As the above discussion indicates, the choice of chlorination control strategy must be dictated by the physical conditions at the WWTP. The use of an on-line residual feedback control strategy is preferable in all but the smallest WWTPs because this will allow disinfection to be accomplished in the most efficient and cost-effective manner. As the concern about chlorine discharge to the environment increases, optimization of the chlorination control strategy will become increasingly more important. In addition to keeping the cost of chlorination to a minimum, maintaining the effluent chlorine residual at the lowest concentration needed to achieve the required degree of disinfection will help minimize the cost of subsequent dechlorination.

One situation that can occur with the use of residual control relates to the breakpoint chlorination reaction. During the breakpoint reaction, an increase in the chlorine dosage will actually lead to a reduced chlorine residual as the available chlorine is consumed in the conversion of ammonia to nitrogen gas. This typically can occur when there is inadequate mixing, with a resultant increase in side reactions such as breakpoint chlorination. This condition can be corrected by improving the mixing regime. For further information on chlorine and chlorination, the reader is directed to White's *Handbook of Chlorination* (1986).

*D*ECHLORINATION CONTROL

Chlorination of WWTP effluents to minimize bacterial contamination of receiving waters has been widely practiced for more than 30 years (White, 1986). More recently, toxic effects of chlorine residuals on aquatic wildlife have been established, and requirements for dechlorination of WWTP discharges are becoming increasingly common (Chen and Gan, 1981; Riley, 1989; and White, 1986).

Current methods of dechlorination include chemical addition, granular activated carbon adsorption, and degradation in holding ponds. Chemicals used to remove chlorine and chloramine compounds include sulfur compounds such as sulfur dioxide, sodium sulfite, sodium bisulfite, and sodium thiosulfate; hydrogen peroxide; and ammonia (Riley, 1989, and White, 1986). Of all these methods, application of sulfur dioxide is the most widely used method in municipal WWTPs, primarily because of the relatively low cost and wide availability of supplies of the chemical (Chen and Gan, 1981). In recent years, the use of sodium bisulfite has been increasing because the capital cost of feed equipment is lower and the safety concerns are significantly reduced when compared to those for sulfur dioxide use. The strategies presented below assume the use of sulfur dioxide. However, much of the discussion below applies to other chemical addition methods, particularly the use of sodium bisulfite and other sulfur compounds.

Proper control of the sulfur dioxide process is important to prevent damage to aquatic life, to avoid discharge violations, to minimize chemical costs, and to avoid undesirable negative publicity. Sulfur dioxide overdosing has not been found to be significantly harmful to receiving waters (Chen and Gan, 1981). However, excess sulfur dioxide addition to lightly buffered effluents can depress the pH, which can violate discharge requirements.

DIFFICULTIES OF DECHLORINATION CONTROL. Dechlorination control can be one of the more difficult processes to automate successfully. The primary reasons include the following:

- The process variable being controlled is difficult to measure.
- Almost every part of the control system and process contain inherent response-time delays.
- The reliability of the control system must be high. Almost any failure can result in a discharge violation.
- Sulfur dioxide (when used) is caustic and toxic, requiring special safety precautions and equipment.
- The desired range of acceptable control is relatively narrow considering the process dynamics and measuring system limitations.
- Every application is unique, requiring careful consideration.

Measurement Limitations and Problems. The fundamental measurement problem of dechlorination control is that the variable being controlled is difficult to measure, particularly in the range at which the variable is controlled. The ideal value for the variable being controlled is zero. However, to control chlorine (Cl_2) residual at zero, it is necessary to be able to measure it in a range around zero, both positive and negative. However, chlorine residual cannot be readily measured in the negative. It is difficult to quantitatively measure the absence of something.

This problem can be illustrated by considering Figure 14.5. This chart shows a scale for chlorine residual on the left and a scale for its neutralizing chemical, sulfur dioxide, on the right. The two scales are aligned to represent the stoichiometry between chlorine residual and sulfur dioxide residual. In the center is the ideal condition: zero chlorine and zero sulfur dioxide. This chart has four quadrants. The upper left quadrant can be measured using a chlorine residual analyzer. The lower right quadrant can be measured using (relatively new) continuous sulfite ion analyzers.

Historically, there have been five basic approaches to the problem of measuring and controlling chlorine residual:

- No on-line residual measurement, with process control consisting of grab samples and laboratory analysis;

- On-line chlorine residual measurement after (partial) dechlorination using feedback control with a positive setpoint near zero;
- Chlorine residual measurement of process stream before sulfur dioxide injection with feedforward control;
- "Biased" chlorine residual measurement after dechlorination using on-line analyzers, generally using feedback control; and
- On-line sulfur dioxide (sulfite ion) residual analyzer, typically combined with on-line chlorine residual analyzer, to measure residual after dechlorination, generally using feedback control.

A feedback control system for dechlorination ideally would have a setpoint at or near zero and measure chlorine and sulfur dioxide residuals over some range of residuals, for example from $+5.0$ mg/L Cl_2 to $+4.5$ mg/L SO_2 (-5.0 mg/L Cl_2). In practice, this has been difficult to achieve. Chlorine residual analyzers are frequently described as being the "weakest link in a sulfur dioxide feed system" (Chen and Gan, 1981).

Analyzer types can be generally classified as amperometric, polarographic, calorimetric, or potentiometric. Another characteristic is that some

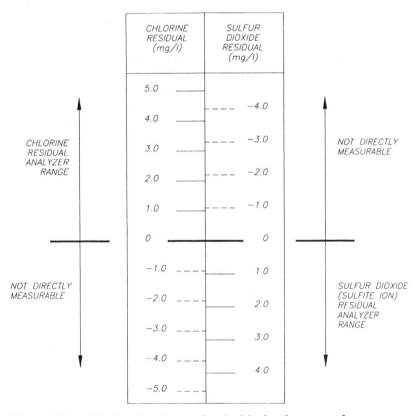

Figure 14.5   Limits of analyzers for dechlorination control

are of the *probe* type; that is, they use an electrode surrounded by an electrolyte solution sealed behind a chlorine-permeable membrane. Others are of the *bare-electrode* type, in which the electrodes come in direct contact with the sample.

Care must be taken in selecting a chlorine residual analyzer. The analyzer must be capable of measuring the species of chlorine present in the effluent stream. Some of the analyzers do not hold calibration when there is a change in the relative concentrations of the species of chlorine. Also, other chemicals sometimes found in wastewater effluents can create interferences to the residual measurement.

The analyzers available have additional limitations. Many chlorine residual analyzers cannot continuously measure residuals less than 0.2 mg/L without severe calibration drift caused by electrode polarization, although some manufacturers of newer products claim they have eliminated the polarization problem.

Some probe-type analyzer designs allow installation directly in the process stream (*in situ*). Considerable care must be used when applying these probes in *in situ* applications because the residual measurement of membrane-covered electrodes is dependent on the sample velocity across the membrane.

To allow measurement over a range spanning both sides of zero residual, a few biased sampling and analysis systems have been developed. These systems so far have not attained widespread acceptance. A method to reliably measure the sulfur dioxide residual downstream of a sulfur dioxide diffuser would greatly aid in dechlorination control. Theoretically, a chlorine residual analyzer and a sulfite ion analyzer could be combined to measure the desired range ($+5$ to -5 mg/L Cl_2). This has not been a widely used method because two different analyzers are required for a single measurement, and sulfite ion analyzers have not been particularly successful in wastewater dechlorination applications.

All of the chlorine residual and sulfur dioxide analyzers can require considerable maintenance. Analyzer maintenance varies from less than 1 hour per week per unit to more than 10 hours per week per unit. One particular maintenance problem in dechlorination applications is biofouling. Biological growth occurs rapidly in dechlorinated effluents and in associated sample lines, in wet chemistry analyzer parts, and in probe-type analyzer membranes. Thus, any equipment being used to measure the dechlorinated effluent requires frequent cleaning.

Time Delays. Figure 14.6 illustrates a typical compound control dechlorination system. Virtually every system element except the feedback controller and electronic signal transmission can introduce a significant time delay that causes the injection of sulfur dioxide to be out of phase with the process requirements.

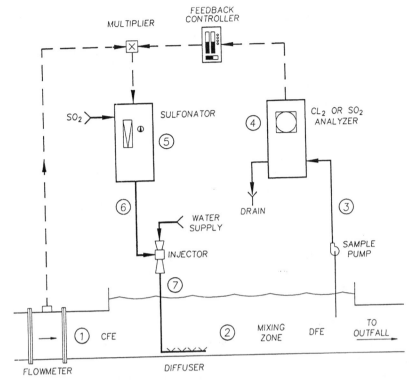

Figure 14.6   Typical control response delays (CFE = chlorinated final effluent and DFE = dechlorinated final effluent)

The following numbered items correspond to the circled numbers in the figure:

1.   The flow meter used for flow-pacing dechlorination is often located distant from the sulfur dioxide diffuser—generally before it, but sometimes after. Any open channels, tanks, weirs, or other structures between the flow meter and the diffuser will tend to cause the flow rate measured by the meter to be different from the flow rate at the diffuser because of hydraulic dynamics. Addition or removal of flow (for example, for filter backwash or reclaimed water) between the flow meter and diffuser will also cause problems.

2.   After sulfur dioxide application, mixing is generally required before final effluent sampling. The time between application and sampling typically can vary between 30 seconds and several minutes.

3.   Time is required to pump the sample to the analyzers unless an *in situ* probe is used. Typical delays are 0.5 to 2 minutes. Biased measurement systems with constant-head tanks have even longer delays.

4. Residual analyzers do not respond instantaneously to sample residual concentration changes. Typical 90% response times to a step change can vary between 5 seconds and 3 minutes depending on the analyzer, the size and direction of the change, and the initial concentration.

5. The valve positioner on a sulfur dioxide feed unit will take 5 to 10 seconds to respond to a 20%-of-range step change in the command input.

6. Because sulfur dioxide gas is a compressible fluid, there is a delay between changes in sulfur dioxide gas feed rate at the sulfonator and at the injector. Depending on the distance, delays between the sulfonator and injector can be from a few seconds to several minutes or more. In some WWTPs, this delay is greater than 15 minutes.

7. There is a short delay from injection of sulfur dioxide into solution to diffusion of the solution into the effluent stream, but this is typically less than 10 seconds.

Note that if a chlorine analyzer measuring the effluent residual before dechlorination is used, it can be subject to timing problems similar to those described in items 1, 3, and 4 above.

Feedforward elements such as the flow meter shown in Figure 14.6 can produce a signal that actually represents a future value at the diffuser. This is called a *leading signal*. A leading signal can cause the same type of control problems as delayed, or *lagging,* signals.

Reliability. Few control loops in WWTPs can result in a virtually instantaneous discharge violation, but dechlorination system failures can and do. Therefore, the reliability required of dechlorination control components is considerably greater than that required of other control loops. Redundant equipment and regular maintenance are necessary, greatly increasing the cost of designs that require multiple analyzers, multiple feed units, or other additional complexity.

Safety Requirements. Sulfur dioxide is a toxic and caustic chemical. Therefore, special design and operation requirements and precautions are employed for storage and delivery systems. These special provisions can create difficulties for dechlorination control. Some WWTPs (particularly those using long outfalls as chlorine contact basins) have located their sulfonators more than 300 m (1 000 ft) from the injectors and diffusers. The purpose is to keep most equipment and the pressure portion of these systems close to the main treatment facility, where leaks can be better contained and controlled. Injectors and diffusers were located near the end of the outfall after chlorine contact. This creates a problematic delay between changes in sulfur dioxide feed rate at the sulfonators and those at the injectors and diffuser.

General-purpose gas and liquid storage and feed equipment is generally avoided in favor of special products from manufacturers specializing in chlorine and sulfur dioxide handling systems. Innovations in sulfur dioxide dechlorination facility design are hampered by an appropriately conservative attitude toward changes in design concepts and by the practice of only using equipment from relatively few specialty manufacturers.

Narrow Control Range Process Dynamics. Most process control loops can vary ± 5 to $\pm 20\%$ around their setpoints without significant economic or regulatory consequences. Excess sulfur dioxide dosage is a costly waste, and any shortage is generally a discharge violation. Because of the negative consequences of chlorine discharge violations, there is a tendency to err on the side of caution and dechlorinate by 1 to 3 mg/L sulfur dioxide more than is required to neutralize the chlorine. This can be a significant additional expenditure at larger WWTPs. In some cases, excess sulfur dioxide in the effluent can cause discharge violations (such as low pH, oxygen depletion, or failed biomonitoring tests), but generally, excess sulfur dioxide in moderate amounts is preferable to having any chlorine in the discharge.

A narrow acceptable control range would not be such a significant problem if the process were operated at nearly steady state. However, most WWTPs not only have significant diurnal fluctuations in flow, but are configured in such a way that certain process changes cause rapid changes in effluent flow. For example, at many WWTPs with tertiary filtration, flow for filter backwashing is withdrawn from between the tertiary filters and the chlorine contact basin. When this occurs, it quickly reduces the flow rate at the postchlorination diffusers, which causes a spike in chlorine concentration, followed by a dip when backwashing is stopped. These chlorine concentration gradients then travel through the contact basin and, if sufficiently large, disrupt the dechlorination process.

Application-Specific Requirements. Many wastewater automatic control schemes are virtually identical from facility to facility. Dechlorination control, conversely, is highly dependent on a number of varying factors. These include the following:

- The performance of residual analyzers is greatly affected by sample physical and chemical characteristics. An analyzer that works well at one WWTP may not work well at another.
- Wastewater treatment plant staff capabilities vary, which can affect their ability to maintain the control system.

- Effluent flow rate characteristics and the physical configuration of tanks, channels, and pipes near the sulfur dioxide diffuser vary, affecting process dynamics.
- The perceived importance of minimizing sulfur dioxide overdosing and minimizing discharge violations varies, thus affecting the determinations by staff to make the control system work properly.
- Economic and safety considerations may require locating the sulfonators a considerable distance from the sulfur dioxide injectors.

Therefore, each design is unique and requires a carefully formulated design.

REVIEW OF CONTROL SCHEMES. Following is a discussion of four types of control loops and three methods for measuring dechlorinated effluent residual. The characteristics, advantages, and disadvantages of each are discussed.

Manual Control. Manual control of dechlorination continues to be practiced at small WWTPs where the cost associated with sulfur dioxide overdosage is relatively small compared to the installation, training, and maintenance costs of an automated dechlorination system. A simplified schematic of a manual dechlorination system is illustrated in Figure 14.7. In a manual system, all sulfur dioxide feed rate adjustments are made by a human operator.

The advantages of this method include

- Low capital, training, and maintenance costs, and
- Simplicity.

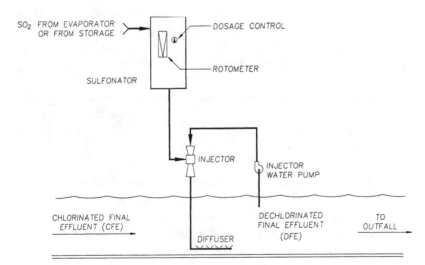

Figure 14.7   Manual control of dechlorination

Disadvantages include

- Requires greater sulfur dioxide overdosage to compensate for process fluctuations,
- Typically requires more frequent grab sampling and testing of final effluent for chlorine residual, and
- Sometimes unacceptable to regulatory authorities because there is no on-line determination of final effluent chlorine residual.

Manual control is used by many WWTPs as the method of backup control, should the automatic control system fail or require maintenance.

Feedforward Control. Feedforward is probably one of the oldest and the most widely used dechlorination control schemes. Its primary advantages include fast response time and avoidance of the problems associated with continuous monitoring of dechlorinated final effluent (DFE).

A typical feedforward control system is shown in Figure 14.8. Flow-residual proportional control is illustrated because it is the most common configuration in systems without feedback control. In this system, the effluent flow rate is multiplied by the chlorine residual upstream of the sulfur dioxide diffuser to calculate the quantity of chlorine that must be neutral-

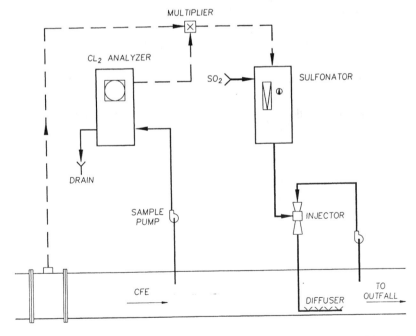

Figure 14.8   Flow-residual feedforward control (CFE = chlorinated final effluent)

ized. Appropriate conversion constants and a safety margin are, in effect, built into the multiplier. The sulfonator output is regulated to provide enough sulfur dioxide to neutralize the chlorine in the effluent, plus a small excess of sulfur dioxide to allow for measurement inaccuracies and time delays. Other forms of feedforward control are sometimes used, with simple flow-proportional control (flow pacing) being the most common.

The advantages of feedforward control include the following:

- It has a fast dynamic response.
- It avoids problems associated with feedback control.
- When properly applied, it is based on a relatively accurate model of the process, so if measurement and final control element errors are minimized, it provides relatively good control.
- It is relatively simple and easy to understand.

Disadvantages include

- No continuous monitoring of dechlorinated effluent residuals. Thus, there is no feedback for the control system to adjust for changes in the sulfur dioxide-to-chlorine dosage ratio.
- No continuous monitoring (and recording) of performance, which is often required by regulatory agencies, especially in California.

Limitations include the following:

- The flow meter used for flow pacing must be properly located to limit differences between the measured flow and the flow at the sulfur dioxide diffusers to less than 10% of the measured rate (including meter error). Because of economic and hydraulic considerations, this consideration is often not included in the WWTP design.
- This system assumes the output of the sulfonator(s) is linear and proportional to the control signal. Often, this is not completely true.

Feedback and Compound Control. As discussed previously, feedback control for dechlorination suffers from an inherent weakness: the time delays present in a feedback dechlorination control system. Therefore, most dechlorination control systems using feedback control also use feedforward control. *Compound control systems* combine feedforward and feedback control. The feedback portion is typically designed to respond at a considerably slower rate than the feedforward portion; therefore, the feedback portion is often referred to as the "trim" system.

A problem with feedback systems is that they require the use of either a biased measurement system, a sulfite ion analyzer, or a chlorine residual analyzer measuring at or near zero. Each of these design options has problems that have been discussed above or will be discussed below.

An example compound control system is illustrated in Figure 14.9. It is basically a feedforward control system with the command signal to the sulfonator multiplied (trimmed) by the output of a feedback controller using DFE residual as its process variable.

Advantages of this system include

- Theoretically, an almost ideal solution: responds quickly to most significant and dynamic process inputs and adjusts via feedback to changes in slower but difficult-to-measure process inputs.
- Continuous monitoring of DFE residual.

Disadvantages include

- Disadvantages of DFE measuring systems discussed below under Dechlorinated Final Effluent Residual Measurement.
- Relatively complex: can be difficult for operators to understand and for technicians to troubleshoot.

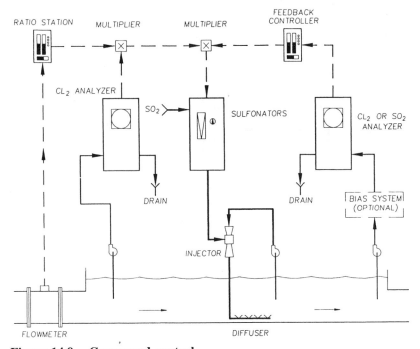

Figure 14.9   Compound control

Two-Step Control. This control scheme is a hybrid system like compound control but does not require measurement of DFE residuals. An example system is illustrated in Figure 14.10. This system uses two points of sulfur dioxide injection and measures the chlorine residual in the section of channel or pipe between the two injection points. It also requires an effluent flow meter.

The sulfonator for the first diffuser is controlled by a compound system: flow-feedforward and residual-feedback. The setpoint for chlorine residual downstream of the first diffuser is set at a positive value, typically 0.3 to 1.0 mg/L. Thus, the chlorine residual analyzer always operates in a positive range high enough to avoid polarization and calibration drift.

The second sulfonator is operated on the feedforward principle, using effluent flow and the same chlorine residual signal used for feedback on the first control loop. Because the second loop neutralizes less than 1.0 mg/L chlorine residual, it is easy to achieve zero residual reliably with relatively little sulfur dioxide overdose. For example, consider a feedforward loop neutralizing 0.5 mg/L chlorine. A 20% positive chlorine residual measurement error would result in a 0.1-mg/L additional overdose.

If, however, the loop were neutralizing 5.0 mg/L chlorine, a positive 20% residual error would result in a 1.0-mg/L overdose. Also, a 20% negative error results in a roughly 1-mg/L underdose.

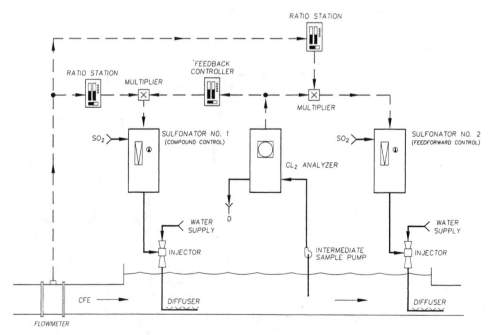

Figure 14.10   Two-step control (CFE = chlorinated final effluent)

Advantages of this scheme include the following:

- Avoids problems associated with measuring DFE chlorine residual.
- Capable of achieving zero residual with relatively little sulfur dioxide overdosage.
- Generally overcomes the limitations of feedforward control associated with inaccuracies and nonlinearities in flow measurement, chlorine residual measurement, and sulfur dispensing.

Disadvantages include the following:

- Does not provide continuous monitoring of DFE.
- Requires two sulfonation systems, each including sulfonators, injectors, and diffusers.
- Relatively complex; may be difficult for operators to understand.

As suggested by White (1986), use of a second sulfonation system might be avoided by splitting the sulfur dioxide solution line, and metering and controlling the line to the second diffuser.

Dechlorinated Final Effluent Residual Measurement. This subject is treated separately from the control schemes described above because there are several alternatives available, and any one can be used with feedback (or compound) control systems.

If feedback control is perceived to be necessary to achieve adequate control without excessive sulfur dioxide overdosage, or if continuous monitoring of the DFE residual is required by the applicable regulatory agency, there are presently three alternatives:

- Use of a stand-alone chlorine residual analyzer;
- Use of a biased measurement system; or
- Use of one of the relatively new sulfur dioxide residual analyzers.

STAND-ALONE CHLORINE RESIDUAL ANALYZER. The first alternative uses a single chlorine analyzer measuring at or near zero. If discharge requirements allow, some WWTPs try to control the dechlorination process to a small positive chlorine residual setpoint, such as 0.1 mg/L. This alternative has been attempted periodically since the early 1970s, generally without success.

Most chlorine residual analyzers rapidly drift out of calibration when measuring a sample stream with little (less than 0.2 mg/L) or no chlorine residual. Also, if there are rapid changes in effluent flow or other fast-changing process dynamics, the typical allowable control ranges of 0.2 to 0.0 mg/L or 0.1 to 0.0 mg/L chlorine residual are too narrow to be maintained.

BIASED MEASUREMENT SYSTEM. There are at least two types of biased measurement system: constant and variable. Many constant bias systems were designed in the 1970s and early 1980s but seem to have fallen out of use. Variable bias systems are based on the concept that if two samples of wastewater are mixed in a 50:50 ratio, with the chlorine residual of one being known, then by measuring the chlorine residual of the resultant mixture, the unknown sample chlorine residual can be calculated. In fact, if the unknown sample contains sulfur dioxide residual, its capacity to neutralize chlorine, represented as negative chlorine residual, can be calculated.

The system illustrated in Figure 14.11 uses constant-head tanks to control and mix the sample flows; other equipment arrangements can be used. Chlorinated final effluent (CFE) and DFE samples are pumped to constant-head tanks. A portion of the CFE sample is also routed directly to one of

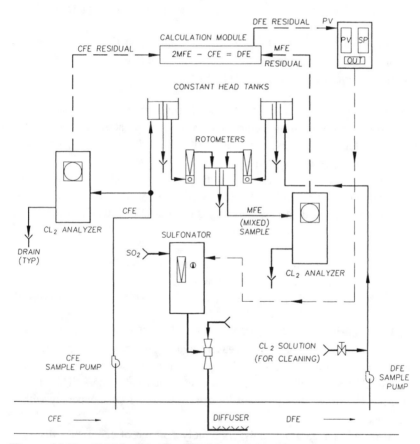

Figure 14.11   Residual feedback with variable bias (shown without feedforward control) (CFE = chlorinated final effluent; DFE = dechlorinated final effluent; and MFE = mixed final effluent)

the two analyzers. The constant-head tanks are designed to be geometrically and hydraulically identical, thus resulting in equal flow from each. The two flow rates are routed through adjustable rotameters (optional) for verification and adjustment of flows and then mixed in a third constant-head tank. Flow from the third tank, mixed final effluent (MFE), is then routed to the second analyzer. The two residual measurements, CFE and MFE, are then used to compute DFE residual, using the formula:

$$DFE = 2MFE - CFE \qquad (14.1)$$

The computed DFE is then used as the process variable in a conventional feedback control loop.

The advantages of this technique include the following:

- Permits feedback control with a process variable whose range is optimal for the application, typically a range of $+5$ mg/L to -5 mg/L chlorine residual;
- Permits continuous monitoring of dechlorination permit compliance and system performance; and
- Potentially can reduce sulfur dioxide overdose.

Disadvantages are as follows:

- A minimum of two, and typically three, chlorine residual analyzers is required (third unit is backup for reliability). This increases calibration and maintenance requirements.
- The system is complex, and its operation may be difficult for operators to understand clearly.
- The large number of components and the complex sample routing increase cleaning and other maintenance requirements.
- The DFE sample system is subject to rapid biofouling. Biofouling can be controlled by daily dosing of the sample lines and analyzers with chlorine for 5 to 10 minutes, but this increases maintenance requirements.

Two additional limitations are the following:

- The system optimally requires a minimum CFE chlorine residual concentration of 2 mg/L to minimize measurement errors. A minimum of 3 mg/L is preferable.
- A feedback control strategy for dechlorination (including use of this method for measuring the DFE residual) cannot maintain control with fast process dynamics because of the response time delay problems discussed above. The proper application of feedforward control

is required to keep up with fast process changes. Thus, biased measurement systems generally must be combined with feedforward control, increasing system complexity.

SULFUR DIOXIDE RESIDUAL ANALYZER. The third alternative for DFE residual measurement is to measure the sulfur dioxide (sulfite ion) residual in the DFE. This alternative has recently become feasible, although it is not yet widely used. In addition to being unproven, sulfur dioxide residual analyzers have a limitation when compared to bias system feedback control: the range of measurement is essentially limited to positive values of sulfur dioxide residual. Optimal feedback control requires the measurement of negative sulfur dioxide residuals (positive chlorine residual) up to at least 2 mg/L. This is necessary so that the corrective action will be proportional to the deviation from the setpoint whenever the sulfur dioxide dosage does not neutralize all of the chlorine.

RECOMMENDATIONS. The following considerations are suggested regarding implementation of dechlorination control systems.

1. Because the chlorine analyzer is still the weak link, choose the analyzer carefully, preferably by testing two or three manufacturers' products in the intended application.
2. Use an accurate, reliable flow meter as close as possible to the point of sulfur dioxide injection. Any lag between the flow measurement and the actual flow at the point of injection will negatively affect the system's ability to control.
3. Control postchlorination to minimize chlorine residual variations through the contact basin. Changes in chlorine residual concentration immediately upstream of the sulfur dioxide injection point are difficult to track. Chlorine contact basins are generally designed to be plug flow, so effluent chlorine residual will reflect significant variations in the quality of the flow to the contact basin.
4. Mixing of the dechlorination chemical into the effluent stream is a critical factor in the performance of the control strategy. The more complete and instantaneous the mixing, the better the system response will be, and the more closely the dosing will approach stoichiometric ratios.
5. The choice of whether to install a biased measurement system to measure dechlorinated effluent chlorine residual and sulfur dioxide residual depends on the size of the WWTP, the sophistication of the staff, the regulatory requirements, and other factors. Biased measurement systems can be useful but are not the most cost-effective solutions for all users.

*O*ZONE DISINFECTION

Ozone has been used in large-scale drinking water disinfection since the early 1900s in Europe. Its use in the U.S. has been mainly in chemical manufacturing. Disinfection of wastewater effluent by ozone is gaining acceptance over chlorine because of its efficacy, the lack of additional dechlorination facilities required, the absence of undesirable byproducts such as halogenated hydrocarbons or residual chemicals toxic to aquatic life, and the benefits of reaeration of the receiving stream resulting from the rapid breakdown of ozone to oxygen.

Ozone is second only to fluorine in oxidizing power. It is a much stronger disinfectant than chlorine and it is quite toxic and unstable. Viruses, bacteria, and other pathogens are attacked much more rapidly by ozone than by chlorine. Because of the relative instability of ozone, it cannot be stored and therefore must be generated on site. Ozone use is relatively safe because it is only produced in low concentrations that are far below explosive levels and are not dangerous when general precautions are taken. Ozone generation most commonly is accomplished by a continuous controlled electrical (corona) discharge in the presence of oxygen. Optimal production efficiency produces ozone concentrations in a range of 2 to 4%. This section presents the operating principles, objectives, and control strategies for the ozone generation and disinfection unit process used in wastewater treatment.

BASIC OPERATING PRINCIPLES. Oxygen gas is typically produced on site by cryogenic air separation or by other methods such as pressure swing absorption, or it is transported to the WWTP site. The oxygen supply may serve both the ozone disinfection process and the activated-sludge facility. To generate ozone, the oxygen pressure is increased by feed compressors and it is fed into the ozone generators. As the oxygen gas passes through the ozone generators, an alternating voltage is imposed across discharge gaps. A uniform corona discharge is created in the discharge gap by the presence of a dielectric material between the high-voltage and ground electrodes. A mixture of gases consisting of ozone in oxygen is produced, with the ozone totaling approximately 2% by weight or greater. The ozone-enriched oxygen gas mixture then flows to the ozone contact tank.

Clarified and/or filtered wastewater flows to the ozone disinfection contact tank. As the wastewater flows through the tank, ozone-enriched oxygen gas is typically sparged into the wastewater through porous stone diffusers. Complete mixing is important for disinfection efficiency. As the gas rises to the liquid surface, practically all of the ozone reacts, dissolves, and decomposes back to oxygen as it disinfects the wastewater. At the discharge end of the contact tank, the disinfected wastewater is typically separated from the gases by an underflow weir system and flows into a contact tank effluent chamber. From the effluent chamber, the wastewater overflows a weir and flows to the WWTP discharge.

Vapors formed in the ozone contact tank effluent chamber are exhausted through a destructor that uses a platinum-based catalyst. The vapors are vented to the atmosphere or mixed with the gases collected under the ozone contact tank cover, which consist chiefly of oxygen; they are then reused in the activated-sludge facility (integrated option) or in the disinfection system (recycle option).

SPECIAL PROCESS CONSIDERATIONS. Ozone is produced when high-voltage alternating current (ac) is imposed across a discharge gap in the presence of an oxygen-containing gas. However, only a fraction (15%) of the energy supplied is used to make ozone. The remainder is lost as light, sound, and, primarily, heat. Unless this heat is removed efficiently, the ozonator gap acts as an oven, and high temperatures build up in the discharge space and at the dielectric surfaces. If temperatures are allowed to build up, ozone yield will suffer because decomposition of ozone is temperature-sensitive, and the dielectric characteristics can be affected to the point of causing failure. Therefore, the monitoring of temperature is important, and an efficient method of heat removal is essential. Cooling systems using water and/or special heat-transfer fluids, such as freon or oils, are used to cool electrodes and electric power conversion equipment.

Moisture must be eliminated because it promotes ozone decomposition. This is typically not a problem when highly pure oxygen, such as is produced in many cryogenic generators, is used because it is already dry. However, the air feed valve for feed compressor start-up may be a source of moisture if it is not properly seated.

All dusts, aerosols, and related compounds must be removed from the gas before its entry into the ionizing chamber. Such pollutants will tend to change the dielectric constant of the gap. Pure oxygen contains no such contaminants.

PROCESS CONTROL STRATEGIES. Many ozonation systems operate in either integrated or recycle modes. With the integrated system, oxygen generated is first used by ozone disinfection, and the ozone-containing vapors from the ozone contact tank (or *offgases*) pass through a catalytic ozone destruction unit before being used for activated-sludge treatment. The recycle option involves the reuse of contact tank offgas for generator feed gas. In both options, the ozone-production system uses variable feed flow or recycle compressors and adjustable power to the ozone generators and ancillary subsystems. The control strategy controls the ozone generator output and the gas flow, which yields total ozone production.

The control strategy determines the ozone production to obtain a desirable residual ozone or offgas concentration equal to the setpoint. As a first approximation, the ozone production is proportional to the wastewater flow. Any significant variation in wastewater flow should be immediately taken into account to minimize the effect on residual ozone.

The main process information inputs are

- Wastewater flow rate,
- Mass flow of oxygen or air to the ozone generators (see equation 14.2),
- Number of generators running, and
- Ozone concentration in the water or concentration in the contactor offgas.

Two means of feedback measurement exist to control the ozone generator output: measurement of ozone in the offgas or measurement of the ozone residual in the effluent. The feedback value is compared with a preset ozone concentration setpoint. The difference trims the control loop. The total ozone production required, as a function of influent flow, quality, and desired ozone residual, divided by the optimum ozone production per generator determines the number of generators that should be in service at a particular time. If there is a discrepancy between the number of generators required and the number running, the operator should be notified that a generator or generators should be started or stopped, if in manual control. Integration of the run time for each generator and the maintenance of a start–stop priority list should determine which generator should be started or stopped to equalize the generator run times and maximize operational availability.

The ozone generator output depends not only on power input but also on the ozone concentration because of recombination effects and oxygen feed purity. Ozone-enriched gas should be produced at constant concentration to optimize generator and contact chamber efficiency. Therefore, it is necessary to control the ozone generator inlet gas flow, inlet gas purity in the case of recycle option, and generator input power as a function of effluent flow and quality.

CONTROL LOOPS. The control of ozone disinfection involves three main control loops: the ozone generator feed gas control loop, the ozone production control loop, and the contact tank pressure control loop. The integrated mode, the usual mode of operation employed, is used in this example and is illustrated in Figure 14.12.

Loop 1: Ozone Generator Feed Gas Flow Control. The gas flow required to maintain the ozone concentration at a constant value is deter-

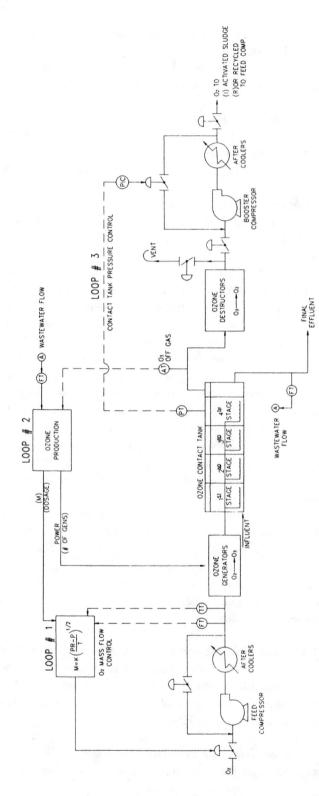

Figure 14.12   Integrated mode of ozone disinfection control

mined from the ozone production or dosage requirements. The flow is obtained by automatic manipulation of the compressor inlet throttle valve or the generator flow valve using a manually entered mass flow setpoint. The mass flow rate signal can be the same as the total output from the oxygen-generating plant, if operating in the integrated mode. Temperature- and pressure-compensated compressor outlet flow measurements are used to continuously calculate a mass feed gas flow rate as shown below

$$M = \frac{K(P_I - P)^{1/2}}{T} \tag{14.2}$$

Where

M = mass flow of gas, kg/m^2·s

P_I = upstream static pressure, N/m^2;

P = differential pressure, N/m^2;

T = inlet gas temperature, K; and

K = proportionality constant, including orifice coefficient and geometry.

Loop 2: Ozone Production. Control of ozone production can be paced from a wastewater flow rate signal along with the ozone offgas concentration. The measured oxygen feed gas mass flow provides input to this loop as well. The desired offgas concentration is manually entered as a setpoint and compared to the actual measured ozone offgas concentration. An algorithm calculates the amount of power required to generate the correct ozone dosage. An output signal is set to an ac frequency controller (not shown) as part of the generator, which determines the actual ozone produced. Ozone mass generation is directly proportional to the ac frequency at a constant voltage.

Loop 3: Contact Tank Pressure Control. The contact tank pressure, with an operator-established setpoint, is controlled by one of two controllers: the vent or booster compressor controller (integrated mode) or a makeup valve controller (recycle mode). In either mode, the setpoint pressure in the enclosed tank is typically maintained slightly negative (that is, − 25 mm [− 1.0 in.] of water column) to prevent incidental leakage of ozone-containing gas to the atmosphere. The booster compressor controls this pressure by throttling. The pressure determines the rate of gas flow from the vapor space of the contact tank through demisters and catalytic ozone destruct units. Once repressurized by these booster compressors, this gas becomes the feed gas to the oxygen activated-sludge system in the integrated mode.

COMPRESSOR SURGE CONTROL. In addition to the above loops, the feed and booster compressors should always be operated with sufficient

---

mass flow to maintain conditions above the surge point. The compressor manufacturer should provide a table or graph of safe operating limits for this purpose. Closed contacts from the antisurge vent or bypass valve used in control should sound an alarm if open to indicate a throttle, vent, or bypass valve or hardware controller malfunction.

PREDICTING OZONE DEMAND. Although the wastewater ozone requirements typically vary relatively slowly, such variations may be considerable in some WWTPs and may necessitate large adjustments in the ozone production rate, in turn affecting power use. Influent ozone requirements (or demand constituents) are measured by biochemical oxygen demand, chemical oxygen demand, turbidity, ammonia concentration, temperature, and other characteristics. The effect of changes in these parameters, if not directly proportional to ozone uptake, may be at least reasonably predictable. The potential exists, therefore, for predicting the production rate of ozone by the gas flow (loop 1) and generator production (loop 2) control loops. This suggests a feedforward strategy using on-line analyzers that provide feedforward inputs. Therefore, some of these parameters measured in the contact tank influent could modify or trim ozone production. As mentioned previously, ozone offgas concentration remains useful as feedback trim as indicated in loop 2 of Figure 14.12. On-line measurement of final effluent ozone concentration may be used instead to pace ozone production and delivery.

OTHER INPUTS. Other on-line measurements in the ozone disinfection process include ambient ozone concentration to detect leaks and oxygen feed gas dew-point temperature to prevent ozone decomposition. Measurement of temperature inputs throughout the generators is important to ensure that the cooling systems are effectively avoiding harmful overheating. Oxygen and hydrocarbon sensors are also useful for monitoring the offgas stream from the contact tank and the ambient air in the ozone generator area for personnel safety.

ULTRAVIOLET DISINFECTION

Ultraviolet disinfection has been used successfully in both small and large WWTPs, avoiding the chemical handling requirements and costs associated with chlorination/dechlorination. The UV disinfection process involves passing wastewater near low-pressure lamps that produce UV light in the region of 250 nanometers. That wavelength is lethal to many microorganisms. The mechanism of microorganism destruction is believed to be the breakage of chemical bonds in the deoxyribonucleic and ribonucleic acids of the organism. Replication is blocked and the cells expire.

The UV disinfection process may be designed as either an open-channel or closed-vessel system. In both systems, the lamps are enclosed in quartz tubes and wastewater passes between the tubes. The spacing between the tubes is in the 25- to 75-mm (1- to 3-in.) range. At large WWTPs, the lamps are separated into sections called *banks* for maintenance and for staging the number of lamps in service at any one time.

The kill is determined by dosage and is a function of the product of intensity and time. The intensity is a function of the number of lamps, the spacing between lamps, the wastewater characteristics, and the degree of scaling on the quartz tubes enclosing the lamps. There is no residual associated with the UV disinfection process, and the dosage is difficult to measure. Feedback of bacterial kill generally takes 24 hours. Therefore, optimizing control of the system is difficult.

CONTROL VARIABLES. The main costs associated with operation of a UV disinfection system are electrical power costs, cleaning of quartz tubes, and replacement of the lamps. Effluent disinfection is dependent on monitoring the primary variables associated with dosage, including flow, degree of scaling, and suspended solids concentration. When the quartz tubes enclosing the lamps are clean, the dose is mainly dependent on effluent suspended solids, effluent transmittance, and the flow rate through the system. With time, a scale builds up on the quartz tubes at a rate that is dependent on the wastewater characteristics. This will cause the dose to decrease with time regardless of the transmittance of the wastewater. At some point in time, the scale will reduce the dosage until disinfection standards cannot be met. The quartz tubes must then be cleaned using acid and/or scouring techniques, and the system intensity can be restored. This cycle is shown graphically in Figure 14.13.

FLOW PACING. Most systems having a capacity of greater than 0.04 m³/s (1 mgd) have automatic control capability for turning on the lamps in banks. As flow to the system increases, additional banks are brought on line. Because UV lamps may cost in the range of $30 to $50 per lamp, excessive starting and stopping of the lamps must be avoided to optimize lamp life unless the lamps are specifically designed for multiple starts and stops. Once a bank of lamps is turned on, that bank should be held on for a minimum time of perhaps several hours, depending on lamp and ballast requirements. Sudden surges of flow should be filtered out, or time delays should be used, to ensure that banks of lamps are not unnecessarily started or stopped. Historical flow data may be used to determine the time of day additional banks should be started.

If there are multiple channels or units, these may be brought on line with increases in flow. With multiple units, flow is typically split equally among the units on line. The level or flow in each unit may be maintained

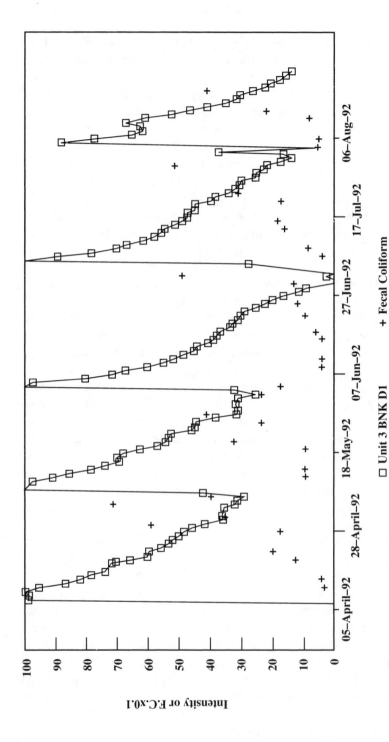

Figure 14.13 Ultraviolet lamp intensity versus time (each peak is a cleaning)

by either a downstream gate, a weir, an adjustable weir, or a valve. An inlet gate may be used to isolate multiple units from a common inlet channel.

The specific control algorithm will depend on the number of units, the hydraulic design, the location of flow meters and level sensors, and the type of downstream control gate or valve. Consideration must be given to ensuring that once a unit is on line, it must be maintained on line for a minimum time period to prevent cycling of the lamp banks. See Chapter 6 for a discussion of flow splitting algorithms.

The control systems for small systems (capacity less than 0.2 m³/s [5 mgd]) need not be overly complex. The control system may consist of a mechanically controlled gate for maintaining individual channel levels and a single flow meter measuring total system flow. Numbers of lamp banks or units may be dependent only on two or three flow ranges, or on the time of day.

For example, a small system may be staged as follows:

- A 0- to 0.02-m³/s (0- to 0.5-mgd) system would consist of one unit and one lamp bank,
- A 0.02- to 0.07-m³/s (0.5- to 1.5-mgd) system would consist of one unit and two lamp banks, and
- A 0.07- to 0.13-m³/s (1.5- to 3.0-mgd) system would consist of two units and four lamp banks.

INTENSITY MONITORING. Most manufacturers of UV systems include intensity photocell monitoring with their systems. These intensity probes can be used to help monitor the efficiency of the UV systems. This information is valuable because a good correlation can be developed between effluent bacterial concentrations and the intensities. Bacterial analyses are required for permit reporting, but these analyses are not good control variables. Bacterial analyses may only be performed one or two times a week, and a 24-hour period must elapse before the results will be available. Continuous indication from the intensity probes can help in making control decisions for the system.

Coupled with suspended solids monitoring and/or absorbance of the effluent, intensity probes can be used to detect the degree of scaling on the quartz tubes. The decrease in intensity shown in Figure 14.13 is not caused by a change in effluent quality but by scaling of the quartz tubes. The data shown in the figure are daily maximum hourly averages for a single-intensity probe.

To date, the intensity probes have not been used for on-line adjustment of flow setpoints for bringing on additional channels or lamp banks. There may be significant fluctuations in the probe readings from minute to minute because of algae or other debris passing in front of, and possibly attaching

---

Disinfection

to, the probe. Alarms on a single probe at any instant in time are probably of limited value and may be a nuisance. Hourly or daily average data for several probes may be used for determining performance and for manually adjusting flow setpoints. The data may also be used for determining when the units should be cleaned to restore full disinfection capacity. In the future, systems may be developed in which the computer may average the values over a period of time, automatically adjust dosage setpoints for turning lamp banks on, and notify the operator when the most cost-effective time for cleaning the units arrives.

LAMP MONITORING. A group of lamps is started and stopped locally by a ballast controlling current to the lamp. A control card typically links the ballast to the lamp. The status of each lamp is monitored by a light-emitting diode (LED) on a local panel or on a control system display. It is important that this panel or display be straightforward and easy to troubleshoot because an indication of a lamp being out may be caused by lamp burnout, ballast failure, control card failure, or failure of a monitoring LED. Monitoring is an important function because as lamps go out, reduced disinfection occurs, and disinfection standards may not be met. Depending on the size of the system, alarming of each individual lamp may not be practical. A limit may be placed on the number of lamps that may be out at any one time, and this may be the alarm setpoint.

R*EFERENCES*

Chen, C., and Gan, H.B. (1981) *Wastewater Dechlorination State-of-the-Art Field Survey and Pilot Studies.* EPA-600/S2-81-169, U.S. EPA, Cincinnati, Ohio.

Riley, P.A. (1989) Wastewater Dechlorination—A Survey of Alternatives. *Public Works.*

Water and Wastewater Instrumentation Testing Association of North America (1990) Performance Evaluation of Residual Chlorine Analyzers For Water and Wastewater Treatment Applications. Report No. CH-1, Washington, D.C.

White, G.C. (1986) *Handbook of Chlorination.* 2nd Ed. Van Nostrand Reinhold Company, New York, N.Y.

Index

A

B

Belt press filtration, 193
  polymer addition, 195

C

Centrifuges, 185
  differential speed control, 191
  feed flow, 187
  instrumentation for control, 189
  machine control, 188
  machine protection and
    maintenance monitoring, 192
  polymer flow, 188
  shutdown sequence, 192
  solid bowl, 186
  start-up sequence, 191
Chlorination control, 200
  compound control loop, 205,
    206
  manual, 202, 203
  on-line monitoring, 204
  residual control, 204
Control
  algorithm, 49
  narratives, 32
  objective, 19
  secondary, 45
  theory, 15
Cost, 6

D

Data logging, 15
Dead time, 12
Dechlorination control, 207
  analyzers, 209
  application-specific
    requirements, 213
  biased measurement system, 220
  compound, 217

dechlorinated final effluent
  residual measurement, 219
difficulties of, 208
feedback and compound , 216
feedforward, 215
flow-residual feedforward, 215
manual, 214
measurement limitations and
  problems, 208
narrow control range process
  dynamics, 213
recommendations, 222
safety requirements, 212
stand-alone chlorine residual
  analyzer, 219
sulfur dioxide residual analyzer,
  222
time delays, 210
two-step, 218
Digital control, 28
Disinfection, 199
Dissolved oxygen, 159
  advanced control, 85
  air drives, 159
  air flow measurement, 88
  biomass inventory control, 160
  blowers, 90
  conventional control, 83
  instrumentation, 86
  mechanical drives, 160
  probe location, 107
  setpoint, 107
Disturbances, 7, 8
Documentation, 31

E

Equilibrium process, 17
Exponential decay, 17

F

Feedback control, 22
Feedforward control, 26

tuning, 25
Pseudolanguage programs, 33
Pump alternation, 53

R

Ratio control, 26
Return activated sludge control,
    115
    fixed flow, 116
    settled sludge method, 118
    sludge blanket level, 117
    solids flux theory, 119
    variable flow, 117
Rotating biological contactor, 157

S

Screen cleaning,
    cumulative flow, 65
    cycles, 65
    differential level, 65
    high influent channel level, 65
Screening and conveyance, 64
    control strategies, 64
Second-order process, 18
Sequencing batch reactors, 141
    continuous influent systems, 142
    equipment, 145
    intermittent influent systems,
        142
Solids waste control, 121
    constant proportion, 133
    control strategies, 135
    decoupling interactions, 131
    downstream effects, 136
    feedback control, 134
    feedforward (ratio) control, 134
    graphical analysis, 130
    hydraulic wasting, 133
    mass balance parameters, 126

mean cell residence time, 128
    process monitoring and
        correlation, 126
    statistical analysis, 130
Standby pumps, 52
Static structures, 57
Steady-state process, 16
Step-feed control, 109
    example, 112
    feedback, 114
    feedforward, 114
    storm flow and bulking, 111

T

Time constants, 18
Trickling filters, 150
    maintenance, 151
    operating, 151

U

Ultraviolet disinfection, 228
    control variables, 229
    flow pacing, 229
    intensity monitoring, 231
    lamp intensity versus time, 230
    lamp monitoring, 232

V

Variable-flow pumps, 50
Variable-speed drives, 51
Variable-speed pumps,
    equalization basin, 53

Z

Ziegler/Nichols, tuning constants,
    12